17

TESI

THESES

tesi di perfezionamento in Matematica sostenuta il 2 novembre 2012

Guido De Philippis
Hausdorff Center for Mathematics
Villa Maria
Endenicher Allee 62
D-53115 Bonn, Germany

*Regularity of Optimal Transport Maps and Applications*

Guido De Philippis

# Regularity of Optimal Transport Maps and Applications

EDIZIONI
DELLA
NORMALE

ISBN 978-88-7642-456-4
e-ISBN 978-88-7642-458-8

# Contents

# Introduction

This thesis is devoted to the regularity of optimal transport maps. We provide new results on this problem and some applications. This is part of the work done by the author during his PhD studies. Other papers written during the PhD studies and not completely related to this topic are summarized in the second part of the introduction.

## 1. Regularity of optimal transport maps and applications

Monge optimal transportation problem goes back to 1781 and it can be stated as follows:

Given two probability densities $\rho_1$ and $\rho_2$ on $\mathbb{R}^n$ (originally representing the height of a pile of soil and the depth of an excavation), let us look for a map $T$ moving $\rho_1$ onto $\rho_2$, *i.e.* such that[1]

$$\int_{T^{-1}(A)} \rho_1(x)\,dx = \int_A \rho_2(y)\,dy \qquad \text{for all Borel sets } A, \qquad (1)$$

and minimizing the total cost of such process:

$$\int c(x, T(x))\rho_1(x)dx = \inf\left\{\int c(x, S(x))\rho_1(x)\,dx \,:\, S \text{ satisfies (1)}\right\}. \quad (2)$$

Here $c(x, y)$ represent the "cost" of moving a unit of mass from $x$ to $y$ (the original Monge's formulation the cost $c(x, y)$ was given by $|x - y|$).

Conditions for the existence of an optimal map $T$ are by now well understood (and summarized without pretending to be aexhaustive in Chapter 1, see [95, Chapter 10] for a more recent account of the theory).

Once the existence of an optimal map has been established a natural question is about its *regularity*. Informally the question can be stated as follows:

*Given two smooth densities, $\rho_1$ and $\rho_2$ supported on good sets, it is true the T is smooth?*

---

[1] From the mathematical point of view we are requiring that $T_\sharp(\rho_1 \mathscr{L}^n) = \rho_2 \mathscr{L}^n$, see Chapter 1.

Or, somehow more precisely, one can investigate how much is the "gain" in regularity from the densities to $T$. As we will see in a moment, a natural guess is that $T$ should have "one derivative" more than $\rho_1$ and $\rho_2$.

To start investigating regularity, notice that (1) can be re-written as

$$|\det \nabla T(x)| = \frac{\rho_1(x)}{\rho_2(T(x)))}, \tag{3}$$

which turns out to be a very degenerate first order PDE. As we already said, the above equation could lead to the guess that $T$ has one derivative more than the densities. Notice however that the above equation is satisfied by *every* map which satisfies (1). Thus, by simple examples, we cannot expect solutions of (3) to be well-behaved. Indeed, consider for instance the case in which $\rho_1 = \mathbf{1}_A$ and $\rho_2 = \mathbf{1}_B$ with $A$ and $B$ smooth open sets. If we right (respectively left) compose $T$ with a map $S$ satisfying $\det \nabla S = 1$ and $S(A) = A$ (resp. $S(B) = B$) we still obtain a solution of (3) which is no more regular than $S$.

It is at this point that condition (2) comes into play. To see how, let us focus on the *quadratic case*, $c(x, y) = |x - y|^2/2$. In this case Brenier Theorem 1.8, ensures that the optimal $T$ is given by the gradient of a convex function, $T = \nabla u$. Plugging this information into (3) we obtain that $u$ solves the following *Monge-Ampère* equation

$$\det \nabla^2 u(x) = \frac{\rho_1(x)}{\rho_2(\nabla u(x)))}. \tag{4}$$

In this way we have obtained a (degenerate) elliptic second order PDE, and there is hope to obtain regularity of $T = \nabla u$ from the regularity of the densities.[2] In spite of the above discussion, also equation (4) it is not enough to ensure regularity of $u$. A simple example is given by the case in which the support of the first density is connected while the support of the second is not. Indeed, since by (1) it follows easily that

$$\overline{T(\operatorname{spt} \rho_1)} = \operatorname{spt} \rho_2,$$

---

[2] One should compare this with the following fact: there is no hope to get regularity of a vector field $v$ satisfying

$$\nabla \cdot v = 0,$$

while if we add the additional condition $v = \nabla u$ we obtain the Laplace equation

$$\Delta v = 0.$$

we immediately see that, even if the densities are smooth on their supports, $T$ has to be discontinuous (cp. Example 1.16). It was noticed by Caffarelli, [21], that the right assumption to be made on the support of $\rho_2$ is *convexity*. In this case any solution of (4) arising from the optimal transportation problem turns out to be a strictly convex *Aleksandrov solution* to the Monge-Ampère equation[3]

$$\det D^2 u = \frac{\rho_1(x)}{\rho_2(\nabla u(x)))} \qquad \text{on Int(spt } \rho_1). \tag{5}$$

As a consequence, under the previous assumptions, we can translate any regularity results for Aleksandrov solutions to the Monge-Ampère equation to solution to the optimal transport problem. In particular, by the theory developed in [18, 19, 20, 89] (see also [66, Chapter 17]) we have the following (see Chapter 2 for a more precise discussion):

- If $\rho_1$ and $\rho_2$ are bounded away from zero and infinity on their support and spt $\rho_2$ is convex, then $u \in C^{1,\alpha}_{\text{loc}}$ (and hence $T \in C^{\alpha}_{\text{loc}}$).

- If, in addition, $\rho_1$ and $\rho_2$ are continuous, then $T \in W^{2,p}_{\text{loc}}$ for every $p \in [1, \infty)$.

- If $\rho_1$ and $\rho_2$ are $C^{k,\beta}$ and, again, spt $\rho_2$ is convex, then $T \in C^{k+2,\beta}_{\text{loc}}$.

A natural question which was left open by the above theory is the Sobolev regularity of $T$ under the only assumptions that $\rho_1$ and $\rho_2$ are bounded away from zero and infinity on their support and spt $\rho_2$ is convex. In [93], Wang shows with a family of counterexamples that the best one can expect is $T \in W^{1,1+\varepsilon}$ with $\varepsilon = \varepsilon(n, \lambda)$, where $\lambda$ is the "pinching" $\| \log(\rho_1/\rho_2(\nabla u))\|_\infty$, see Example 2.21.

Apart from being a very natural question from the PDE point of view, Sobolev regularity of optimal transport maps (or equivalently of Aleksandrov solutions to the Monge-Ampère equation) has a relevant application to the study of the semigeostrophic system, as was pointed out by Ambrosio in [4]. This is a system of equations arising in study of large oceanic and atmospheric flows. Referring to Chapter 5 for a more accurate discussion we recall here that the system can be written, after a

---

[3] This kind of solutions have been introduced by Aleksandrov in the study of the Minkowski Problem: given a function $\kappa : \mathbb{S}^{n-1} \to [0, \infty)$ find a convex body $\mathcal{K}$ such that the Gauss curvature of its boundary is given by $\kappa \circ \nu_{\partial \mathcal{K}}$. All the results of Chapters 2, 3, 4, apply to this problem as well.

suitable change of variable, as

$$
\begin{cases}
\partial_t \nabla P_t + (\boldsymbol{u}_t \cdot \nabla) \nabla P_t = J(\nabla P_t - x) & \text{in } \Omega \times (0, +\infty) \\
\nabla \cdot \boldsymbol{u}_t = 0 & \text{in } \Omega \times (0, +\infty) \\
\boldsymbol{u}_t \cdot \nu_\Omega = 0 & \text{in } \partial\Omega \times (0, +\infty) \\
P_0 = P^0 & \text{in } \Omega,
\end{cases}
\tag{6}
$$

where $\Omega$ is an open, bounded and convex subset of $\mathbb{R}^3$ and

$$
J := \begin{pmatrix} 0 & -1 & 0 \\ 1 & 0 & 0 \\ 0 & 0 & 0 \end{pmatrix}.
$$

We look for solutions $P_t$ which are *convex* for every $t$ (this ansatz is based on the Cullen stability principle [34, Section 3.2]). If we consider the measure[4] $\rho_t = (\nabla u)_\sharp \mathscr{L}^3_\Omega$, then $\rho_t$ solves (formally) the following continuity type equation

$$
\begin{cases}
\partial_t \rho_t + \nabla \cdot (\mathcal{U}_t \rho_t) = 0 \\
\mathcal{U}_t(x) = J(x - \nabla P_t^*(x)) \\
(\nabla P_t^*)_\sharp \rho_t = \mathscr{L}^3_\Omega,
\end{cases}
\tag{7}
$$

where $P_t^*$ is the convex conjugate of $P_t$. Even if the velocity field $\mathcal{U}_t$ is coupled with the density through a highly non-linear equation, existence of (distributional) solutions of (7) can be obtained under very mild conditions on the initial densities $\rho_0 = (\nabla P_0)_\sharp \mathscr{L}^3_\Omega$, [13]. Given a solution of (7) we can formally obtain a solution to (6) by taking $P_t = (P_t^*)^*$ and

$$
\boldsymbol{u}_t(x) := [\partial_t \nabla P_t^*](\nabla P_t(x)) + [\nabla^2 P_t^*](\nabla P_t(x)) J(\nabla P_t(x) - x). \tag{8}
$$

To give a meaning to the above velocity field we have to understand the regularity of $\nabla^2 P_t^*$ where $P_t^*$ satisfies $(\nabla P_t^*)_\sharp \rho_t = \mathscr{L}^3_\Omega$. Notice that the only condition we get for free is that $\mathcal{U}_t$ has zero divergence. In particular, if the initial density $\rho_0$ is bounded away from zero and infinity, the same it is true (with the same bounds) for $\rho_t$. It is then natural to study the $W^{2,1}$ regularity of solutions of (5) under the only assumption that the right hand side is bounded between two positive constants. This is done in

--------

[4] With $\mathscr{L}^3_\Omega$ we denote the normalized Lebesgue measure restricted to $\Omega$:

$$
\mathscr{L}^3_\Omega := \frac{1}{\mathscr{L}^3(\Omega)} \mathscr{L}^3 \llcorner \Omega.
$$

Chapters 3 and 4 (based on [40, 41] in collaboration with Alessio Figalli, and on [44] in collaboration with Alessio Figalli and Ovidiu Savin) while in Chapter 5 (based on [5, 6] in collaboration with Luigi Ambrosio, Maria Colombo and Alessio Figalli) we study the applications of this results to the semigeostrophic system.

Finally we came back to the regularity of solutions of (2) with a general cost function $c$, referring to Section 1.3 for a more complete discussion. In this case, apart from the obstruction given by the geometry of the target domain (as in the quadratic cost case) it has been shown in [80, 78] that a structural condition on the cost function, the so called *MTW-condition*, is needed in order to ensure the smoothness of the optimal transport map. In particular if the above condition does not hold it is possible to construct two smooth densities such that the optimal map between them is even discontinuous.

In spite of this, one can try to understand how large can be the set of discontinuity points of optimal maps between two smooth densities for a generic smooth cost $c$. In Chapter 6 (based on [43] in collaboration with Alessio Figalli), we will show that, under very mild assumptions on the cost $c$ (essentially the one needed in order to get existence of optimal maps), there exist two closed and Lebesgue negligible sets $\Sigma_1$ and $\Sigma_2$ such that

$$T : \operatorname{spt} \rho_1 \setminus \Sigma_1 \to \operatorname{spt} \rho_2 \setminus \Sigma_2$$

is a smooth diffeomorphism. A similar result holds true also in the case of optimal transportation on Riemannian manifolds with cost $c = d^2/2$. Up to now similar results were known only in the case of quadratic cost when the support of the target density is not convex, [52, 55]. We remark here that in this case the obstruction to regularity is given only by the geometry of the domain, while in the case of a generic cost function $c$ we have to face the possible failure of the MTW condition at every point. Thus, to achieve the proof of our result, we have to use a completely different strategy.

We conclude this first part of the introduction with a short summary of each chapter of the thesis (more details are given at the beginning of each chapter):

- **Chapter 1.** In this Chapter we briefly recall the general theory of optimal transportation, with a particular focus on the case of quadratic cost in $\mathbb{R}^n$. We also show how to pass from solutions of the Monge-Ampère equation given by the optimal transportation to Aleksandrov solutions to the Monge-Ampère equation in case the support of the target density is convex. Finally in the last Section we address the case of a general cost function.

- **Chapter 2.** We start the study of the regularity of Aleksandrov solutions to the Monge-Ampère equation, in particular we give a complete proof of Caffarelli's $C^{1,\alpha}$ regularity theorem.
- **Chapter 3.** We start investigating the $W^{2,1}$ regularity of Aleksandrov solutions to the Monge-Ampère equation. We give a complete proof of the results in [40], where we show that $D^2 u \in L \log L$. Then, following the subsequent paper [44], we show how the above estimate can be improved to $D^2 u \in L^{1+\varepsilon}$ for some small $\varepsilon > 0$. We also give a short proof of the above mentioned Caffarelli $W^{2,p}$ estimates.
- **Chapter 4.** Here, following [41], we show the (somehow surprising) stability in the *strong* $W^{2,1}$ topology of Aleksandrov solutions with respect to the $L^1$ convergence of the right-hand sides.
- **Chapter 5.** In this Chapter, based on [5, 6], we apply the results of the previous chapters to show the existence of a distributional solution to the semigeostrophic system (6) in the 2-dimensional periodic case and in the case of a bounded convex 3-dimensional domain $\Omega$. In the latter case we have to impose a suitable decay assumption on the initial density $\rho_0 = (\nabla P_0)_\sharp \mathscr{L}_\Omega^3$.
- **Chapter 6.** Here we report the partial regularity theorems for solutions of the optimal transport problem for a general cost function $c$ proved in [43].

## 2. Other papers

In this second part of the introduction we give a short summary of the additional research made during the PhD studies, only vaguely related to the theme of the thesis. We briefly report the results obtained and we refer to the original papers for a more complete treatment of the problem and the relevant literature.

### 1. $\Gamma$-convergence of non-local perimeter

In [29] Caffarelli-Roquejoffre and Savin introduced the following notion of *non-local* perimeter of a set $E$ relative of an open set $\Omega$:

$$
\mathcal{J}_s(E, \Omega) = \int_{E \cap \Omega} \int_{E^c \cap \Omega} \frac{dxdy}{|x-y|^{n+s}} + \int_{E \cap \Omega} \int_{E^c \cap \Omega^c} \frac{dxdy}{|x-y|^{n+s}}
$$
$$
+ \int_{E \cap \Omega^c} \int_{E^c \cap \Omega} \frac{dxdy}{|x-y|^{n+s}},
$$

and study the regularity of local minimizers of it. This functional naturally arises in the study of phase-transitions with a non-local interaction term, see the nice survey [63] and reference therein for an updated account of the theory.

In [10], in collaboration with Luigi Ambrosio and Luca Martinazzi, we show the $\Gamma$-convergence of the functional $(1-s)\mathcal{J}_s(\cdot, \Omega)$ to the classical De Giorgi perimeter $\omega_{n-1}P(\cdot, \Omega)$ with respect to the topology of locally $L^1$ convergence of sets (a similar earlier result has been obtained in [30] for the convergence of local minimizers of the functionals). We also show equicoercivity of the functionals. More precisely we prove:

**Theorem.** *Let $s_i \uparrow 1$, then the following statements hold:*

(i) *(Equicoercivity). Assume that $E_i$ are measurable sets satisfying*

$$\sup_{i\in\mathbb{N}}(1-s_i)\mathcal{J}_{s_i}^1(E_i, \Omega') < \infty \qquad \forall\Omega' \Subset \Omega.$$

*Then $\{E_i\}_{i\in\mathbb{N}}$ is relatively compact in $L^1_{\mathrm{loc}}(\Omega)$ and any limit point $E$ has locally finite perimeter in $\Omega$.*

(ii) *($\Gamma$-convergence). For every measurable set $E \subset \mathbb{R}^n$ we have*

$$\Gamma - \lim_{s\uparrow 1}(1-s)\mathcal{J}_s(E, \Omega) = \omega_{n-1}P(E, \Omega).$$

*with respect to the the $L^1_{\mathrm{loc}}$ convergence of the corresponding characteristic functions in $\mathbb{R}^n$.*

(iii) *(Convergence of local minimizers). Assume that $E_i$ are local minimizers of $\mathcal{J}_{s_i}(\cdot, \Omega)$, and $E_i \to E$ in $L^1_{\mathrm{loc}}(\mathbb{R}^n)$. Then*

$$\limsup_{i\to\infty}(1-s_i)\mathcal{J}_{s_i}(E_i, \Omega') < +\infty \qquad \forall\Omega' \Subset \Omega,$$

*$E$ is a local minimizer of $P(\cdot, \Omega)$ and $(1-s_i)\mathcal{J}_{s_i}(E_i, \Omega') \to \omega_{n-1}P(E, \Omega')$ whenever $\Omega' \Subset \Omega$ and $P(E, \partial\Omega') = 0$.*

## 2. Sobolev regularity of optimal transport map and differential inclusions

In [9], written in collaboration with Luigi Ambrosio and Bernd Kirchheim, we started the investigation of the Sobolev regularity of (stricly convex) Aleksandrov solution to the Monge-Ampé re equation. More precisely we show that in the 2-dimensional case the Sobolev regularity of optimal transport maps is *equivalent* to the rigidity of a partial differential inclusion for Lipschitz maps (see [74, 84] for nice surveys on partial differential inclusions). Referring to the paper for more details, let us define the set of "admissible" gradients

$$\mathcal{A} := \left\{ M \in \mathrm{Sym}^{2\times 2} : \|M\| \leq 1, \ (\lambda+1)|\mathrm{Trace}(M)| \right.$$
$$\left. \leq (1-\lambda)(1 + \det(M)) \right\}, \tag{9}$$

where $\| \cdot \|$ is the operator norm, and the subset $\mathcal{S}$ of "singular" gradients is defined by

$$\mathcal{S} := \left\{ R^{-1} \begin{pmatrix} 1 & 0 \\ 0 & -1 \end{pmatrix} R : R \in SO(2) \right\}. \tag{10}$$

Our main result says that the following two problems are equivalent

**Problem 1.** *Let $\Omega \subset \mathbb{R}^2$ be a bounded open convex set and let $u : \Omega \to \mathbb{R}$ be a strictly convex Aleksandrov solutions to the Monge-Ampère equation*

$$\lambda \leq \det D^2 u \leq 1/\lambda \qquad in \ \Omega.$$

*Show that $u \in W^{2,1}_{\mathrm{loc}}$.*

**Problem 2.** *Let $B \subset \mathbb{R}^2$ be a connected open set, $f : B \to \mathbb{R}^2$ Lipschitz, and assume that $Df \in A \ \mathscr{L}^2$-a.e. in $B$. Show that if the set*

$$\{x \in B : \ Df(x) \in \mathcal{S}\}$$

*has positive $\mathscr{L}^2$-measure, then $f$ is locally affine.*

At the time we wrote the paper we were not able to solve none of the above problems. Notice that the result of Chapter 3 gives a positive answer to Problem 1. In particular this show (in a very unconventional way) that the inclusion in Problem 2 is rigid.

## 3. A non-autonomous chain rule in $W^{1,p}$ and $BV$

In [7], in collaboration with Luigi Ambrosio, Giovanni Crasta and Virginia De Cicco, we prove a non-autonomous chain-rule in $BV$ when the function with which we left compose has only a $BV$-regularity in the $x$ variable. This type of results have some application in the study of conservation laws and semicontinuity of non-autonomous functionals (again we refer to the original paper for a more complete discussion and the main notation). The main result of [7] is the following:

**Theorem.** *Let $F : \mathbb{R}^n \times \mathbb{R}^h \to \mathbb{R}$ be satisfying:*

(a) $x \mapsto F(x, z)$ *belongs to $BV_{\mathrm{loc}}(\mathbb{R}^n)$ for all $z \in \mathbb{R}^h$;*
(b) $z \mapsto F(x, z)$ *is continuously differentiable in $\mathbb{R}^h$ for almost every $x \in \mathbb{R}^n$.*

*Assume that $F$ satisfies, besides (a) and (b), the following structural assumptions:*

(H1) *For some constant $M$, $|\nabla_z F(x, z)| \leq M$ for all $x \in \mathbb{R}^n \setminus C_F$ and $z \in \mathbb{R}^h$.*

(H2) *For any compact set $H \subset \mathbb{R}^h$ there exists a modulus of continuity $\tilde{\omega}_H$ independent of $x$ such that*

$$|\nabla_z F(x, z) - \nabla_z F(x, z')| \le \tilde{\omega}_H(|z - z'|)$$

*for all $z$, $z' \in H$ and $x \in \mathbb{R}^n \setminus C_F$.*

(H3) *For any compact set $H \subset \mathbb{R}^h$ there exist a positive Radon measure $\lambda_H$ and a modulus of continuity $\omega_H$ such that*

$$|\widetilde{D}_x F(\cdot, z)(A) - \widetilde{D}_x F(\cdot, z')(A)| \le \omega_H(|z - z'|)\lambda_H(A)$$

*for all $z$, $z' \in H$ and $A \subset \mathbb{R}^n$ Borel.*

(H4) *The measure*

$$\sigma := \bigvee_{z \in \mathbb{R}^h} |D_x F(\cdot, z)|,$$

*(where $\bigvee$ denotes the least upper bound in the space of nonnegative Borel measures) is finite on compact sets, i.e. it is a Radon measure.*

*Then there exists a countably $\mathcal{H}^{n-1}$-rectifiable set $\mathcal{N}_F$ such that, for any function $u \in BV_{\mathrm{loc}}(\mathbb{R}^n; \mathbb{R}^h)$, the function $v(x) := F(x, u(x))$ belongs to $BV_{\mathrm{loc}}(\mathbb{R}^n)$ and the following chain rule holds:*

(i) *(diffuse part) $|Dv| \ll \sigma + |Du|$ and, for any Radon measure $\mu$ such that $\sigma + |Du| \ll \mu$, it holds*

$$\frac{d\widetilde{D}v}{d\mu} = \frac{d\widetilde{D}_x F(\cdot, \tilde{u}(x))}{d\mu} + \nabla_z \tilde{F}(x, \tilde{u}(x))\frac{d\widetilde{D}u}{d\mu} \qquad \mu\text{-}a.e. \text{ in } \mathbb{R}^n.$$

(ii) *(jump part) $J_v \subset \mathcal{N}_F \cup J_u$ and, denoting by $u^{\pm}(x)$ and $F^{\pm}(x, z)$ the one-sided traces of $u$ and $F(\cdot, z)$ induced by a suitable orientation of $\mathcal{N}_F \cup J_u$, it holds*

$$D^j v = \left(F^+(x, u^+(x)) - F^-(x, u^-(x))\right)\nu_{\mathcal{N}_F \cup J_u}\mathcal{H}^{n-1} \llcorner (\mathcal{N}_F \cup J_u)$$

*in the sense of measures.*

*Moreover for a.e. $x$ the map $y \mapsto F(y, u(x))$ is approximately differentiable at $x$ and*

$$\nabla v(x) = \nabla_x F(x, u(x)) + \nabla_z F(x, u(x))\nabla u(x) \qquad \mathscr{L}^n\text{-}a.e. \text{ in } \mathbb{R}^n.$$

A similar result holds true also in the Sobolev case.

## 4. Aleksandrov-Bakelman-Pucci estimate for the infinity Laplacian

In [31], with Fernando Charro, Agnese Di Castro and Davi Máximo, we investigate the validity of the classical Aleksandrov-Bakelman-Pucci estimates for the *infinity laplacian*

$$\Delta_\infty u := \left\langle D^2 u \frac{\nabla u}{|\nabla u|}, \frac{\nabla u}{|\nabla u|} \right\rangle.$$

The ABP estimate for a solution of a uniformly elliptic PDE states that

$$\sup_\Omega u \le \sup_{\partial\Omega} u + C(n, \lambda, \Lambda) \operatorname{diam}(\Omega) \|f\|_{L^n(\Omega)}, \tag{11}$$

for $f$ the right-hand side of the equation and $0 < \lambda \le \Lambda$ the ellipticity constants (see for instance [25]). A particular useful feature of the above estimates is the presence of an integral norm on the right hand side. In particular the above estimate plays a key role in the proof of the Krylov-Safonov Harnack inequality for solutions to a non-divergence form elliptic equation (see [25]).

In [31] we show that such an estimate cannot hold for solutions of

$$-\Delta_\infty u = f, \tag{12}$$

at least with the $L^n$ norm of $f$ in the right hand side. However we show that a (much weaker) form of the estimate is avaible, namely

$$\left(\sup_\Omega u - \sup_{\partial\Omega} u^+\right)^2 \le C \operatorname{diam}(\Omega)^2 \int_{\sup_{\partial\Omega} u^+}^{\sup_\Omega u} \|f\|_{L^\infty(\{u=\Gamma_u=r\})} \, dr, \tag{13}$$

where $\Gamma_u$ is the convex envelope of $u$. Even if this estimate is weaker than (11) it is still stronger that the plain $L^\infty$-estimate:

$$\sup_\Omega u \le \sup_{\partial\Omega} u + C(n) \operatorname{diam}(\Omega)^2 \|f\|_{L^\infty(\Omega)}.$$

Moreover we are able to obtain a full family of estimates of the type of (13) for solutions of the non-variational $p$-laplacian equation:

$$-\Delta_p u = f,$$

where

$$\Delta_p u := \frac{1}{p} |\nabla u|^{2-p} \operatorname{div}\left(|\nabla u|^{p-2}\nabla u\right). \tag{14}$$

Using that our estimates are stable as $p$ goes to $+\infty$ and some simple comparison argument we also show that viscosity solutions to (14) converges as $p \to +\infty$ to solutions of (12).

## 5. Stability for the Plateau problem

In [45], together with Francesco Maggi, we study the global stability of smooth solution to the Plateau problem in the framework of Federer and Fleming codimension one integral currents, [49]. More precisely we prove that a global stability inequality is equivalent to its local counterpart, namely the strict positivity of the *shape-operator*. Our main result reads as follows

**Theorem.** *Let M be a smooth* $(n-1)$ *dimensional manifold with boundary which is uniquely mass minimizing as an integral* $n-1$-*current. The two following statements are equivalent:*

(a) *The first eigenvalue* $\lambda(M)$ *of the second variation of the area at M,*

$$\lambda(M) = \inf\left\{\int_M |\nabla^M \varphi|^2 - |\mathrm{II}_M|^2 \varphi^2 \, d\mathcal{H}^n : \varphi \in C_0^1(M), \int_M \varphi^2 \, d\mathcal{H}^n = 1\right\},$$

*is strictly positive.*

(b) *There exists* $\kappa > 0$, *depending on M, such that, if* $M'$ *is a smooth manifold with the same boundary of M, then, for some Borel set* $E \subset \mathbb{R}^n$ *with* $\partial E$ *equivalent up to a* $\mathcal{H}^{n-1}$-*null set to* $M \triangle M'$,

$$\mathcal{H}^{n-1}(M') - \mathcal{H}^{n-1}(M) \geq \kappa \, \min\left\{\mathscr{L}^n(E)^2, \mathscr{L}^n(E)^{(n-1)/n}\right\}.$$

We also obtain similar statements in the case of a particular family of singular minimizing cones.

## 6. Stability for the second eigenvalue of the Stekloff-Laplacian

In [15], together with Lorenzo Brasco and Berardo Ruffini, we address the study of the stability of the following spectral optimization problem

$$\max\left\{\sigma_2(\Omega) : \quad \Omega \subset \mathbb{R}^n \quad |\Omega| = |B_1|\right\}. \tag{15}$$

Here $\sigma_2(\Omega)$ denotes the first non trivial Stekloff eigenvalue of the laplacian, *i.e.*

$$\begin{cases} -\Delta u = 0 & \text{in } \Omega \\ \nabla u \cdot \nu_\Omega = \sigma_2(\Omega)u & \text{on } \partial\Omega, \end{cases}$$

with $u$ not identically constant. In [17, 94] it has been showed that the maximum is achieved by balls. The proof is based on the following isoperimetric property of the ball:

$$P_2(\Omega) \geq P_2(B_1) \quad \forall \Omega : |\Omega| = |B_1|, \tag{16}$$

where

$$P_2(\Omega) := \int_{\partial\Omega} |x|^2.$$

The above isoperimetric type inequality has been proved by Betta, Brock, Mercaldo, Posteraro in [14] through a symmetrization technique.

We enforce (15) in a quantitative way, namely we prove that there exists a positive (and computable) constant $c_n$ such that

$$\sigma_2(\Omega) \le \sigma_2(B)\left(1 - c_n \mathcal{A}^2(\Omega)\right) \quad \forall \Omega : |\Omega| = |B_1| \qquad (17)$$

where we have introduced the asymmetry of $\Omega$

$$\mathcal{A}(\Omega) := \min\left\{\frac{|B \Delta \Omega|}{|B|} \quad B \text{ ball, } |B| = |\Omega|\right\}.$$

To prove (17) we had to show a quantitative version of (16), that reads as

$$P_2(B_1)\left(1 + \tilde{c}_n |\Omega \Delta B_1|^2\right) \le P_2(\Omega) \quad \forall \Omega : |\Omega| = |B_1|. \qquad (18)$$

In order to do this, we give a simpler proof of (16) through calibrations which allows to take care of all the reminders in order to obtain (18).

Showing that (17) is optimal, *i.e.* that there exists a sequence of sets $\Omega_\varepsilon$ converging to $B_1$ such that

$$\sigma_2(\Omega_\varepsilon) - \sigma_2(B_1) \approx \mathcal{A}^2(\Omega_\varepsilon),$$

requires some fine constructions due to the fact the $\sigma_2(B_1)$ is a multiple eigenvalue.

## 7. Regularity of the convex envelope

In [42] with Alessio Figalli we investigate the regularity of the convex envelope of a continuous function $v$ inside a convex domain $\Omega$:

$$\Gamma_v(x) := \sup\{\ell(x) : \ell \le v \text{ in } \overline{\Omega}, \ell \text{ affine}\}.$$

We prove the following two theorems:

**Theorem.** *Let* $\alpha, \beta \in (0, 1]$, $\Omega$ *be a bounded convex domain of class* $C^{1,\beta}$, *and* $v : \overline{\Omega} \to \mathbb{R}$ *be a globally Lipschitz function which is* $(1 + \alpha)$-*semiconcave[5] in* $\overline{\Omega}$. *Then* $\Gamma_v \in C^{1,\min\{\alpha,\beta\}}_{\text{loc}}(\Omega)$.

---

[5] Given $\alpha \in (0, 1]$, a continuous function $v$ is said to be $(1 + \alpha)$-semiconcave in $\overline{\Omega}$ if for every $x_0 \in \overline{\Omega}$ there exists a slope $p_{x_0} \in \mathbb{R}^n$ such that

$$v(x) \le v(x_0) + p_{x_0} \cdot (x - x_0) + C|x - x_0|^{1+\alpha} \quad \text{for every } x \in \overline{\Omega} \cap B(x_0, \varrho_0).$$

for some constants $C$ and $\varrho_0$ independent of $x_0$.

**Theorem.** *Let $\Omega$ be a bounded uniformly convex domain of class $C^{3,1}$, and let $v \in C^{3,1}(\overline{\Omega})$. Then $\Gamma_v \in C^{1,1}(\overline{\Omega})$.*

As we show in the paper, the above results are optimal for what concerns the dependence of the regularity of $\Gamma_v$ both on $v$ and on $\Omega$.

# Chapter 1
# An overview on optimal transportation

Monge optimal transportation problem can be stated as follows: given two (topological) spaces $X$ and $Y$, two (Borel) probability measures $\mu \in \mathcal{P}(X)$, $\nu \in \mathcal{P}(Y)$ and a *cost function* $c : X \times Y \to \mathbb{R}$ we look for a map $T : X \to Y$ such that $T_\sharp \mu = \nu$ [1] and that minimize

$$\int c(x, T(x))d\mu(x) = \inf_{S_\sharp \mu = \nu} \int c(x, S(x)). \qquad (1.1)$$

In general, there could be no solution to the above problem both because the class of admissible maps is empty (for instance in the case in which $\mu$ is a Dirac mass and $\nu$ is not) or because the infimum is not attained (see [95, Example 4.9]).

Nevertheless it has been proved by many authors (see [95, Chapter 10] or [11] for an updated account of the theory and some historical remarks) that: if $X$ and $Y$ are, for instance, open subsets of $\mathbb{R}^n$ or of some Riemannian manifold $M$ and $N$, $\mu$ is absolutely continuous with respect to the volume measure, and $c$ satisfies some structural condition (see Section 1.3), then there exists a (unique) optimal transport map $T$. Moreover, $T$ is related to the gradient of a potential $u$ (see Theorem 1.28 for a precise statement).

The aim of this Chapter is to briefly recall some aspects of this theory. In Section 1.1, we start reviewing with some details the case of the quadratic cost on $\mathbb{R}^n$, $c(x, y) = |x - y|^2/2$. In this case Brenier Theorem (Theorem 1.8) states that the map $T$ is given by the gradient of a convex function, that is $T = \nabla u$ for a convex potential $u$. This leads to the

---

[1] In general given a measure $\mu$ we define its *push forward* through a Borel map $T$ as the measure $T_\sharp \mu = \mu \circ T^{-1}$, that is the only measure such that

$$. \int h(y)dT_\sharp \mu(y) = \int h(T(x))d\mu(x) \quad \forall h \text{ Borel and bounded.}$$

celebrated Brenier Polar Factorization Theorem (Theorem 1.11) which can be thought as a Lagrangian version of the Helmoltz Decomposition Theorem for vector filed (see Remark 1.12 ).

In Section 1.2 we show how the convex potential $u$ satisfies a Monge-Ampère type equation almost everywhere and we investigate if this is sufficient to establish some regularity for the transport map. Simple examples (see Example 1.22) show that in order to obtain regularity one needs to impose some condition on the geometry of spt $v$, the support of the target measure, namely *convexity*. Indeed, under this assumption, Caffarelli proved that $u$ is an *Aleksandrov solution* of the Monge-Ampère equation, [21].

Finally in Section 1.3 we briefly sketch how to adapt the results of Section 1.1 to more general cost functions and we discuss (without proofs) the issue of global regularity of transport maps in this case.

Nice and complete references to the theory of optimal transportation are [11] and [95].

## 1.1. The case of the quadratic cost and Brenier Polar Factorization Theorem

Here we study with some details the case of the quadratic cost, with the exception of Chapter 6, this is the case in which we will be mainly interested, for this reason we give some details of the proofs.

Monge Problem for the quadratic cost can be stated as follows: given $\mu$ and $v$ in $\mathscr{P}(\mathbb{R}^n)$, look for a map $T : \mathbb{R}^n \to \mathbb{R}^n$ such that

$$\int |x - T(x)|^2 d\mu(x) = \inf_{S_\sharp \mu = v} \int |x - S(x)|^2, \qquad (1.2)$$

To start studying problem (1.2), following the ideas of Kantorovich, we introduce its relaxed version

$$\min_{\gamma \in \Gamma(\mu, v)} \int |x - y|^2 d\gamma(x, y), \qquad (1.3)$$

where

$$\Gamma(\mu, v) := \left\{ \gamma \in \mathscr{P}(\mathbb{R}^n \times \mathbb{R}^n) \colon (\pi_1)_\sharp \gamma = \mu \ (\pi_2)_\sharp \gamma = v \right\} \qquad (1.4)$$

is the set of *transport plans* between $\mu$ and $v$ (here $\pi_1$ and $\pi_2$ are, respectively, the projection on the first and second factor).

**Remark 1.1.** Given any transport map $T$, *i.e.* a map such that $T_\sharp \mu = v$, it clearly induces a transport plan $\gamma_T := (\text{Id} \times T)_\sharp \mu$. Moreover it can be easily shown (see [11, Lemma 1.20]) that if a plan $\gamma$ is concentrated[2] on the graph of a function $T$, then $\gamma = (\mathrm{Id} \times T)_\sharp \mu$.

Thanks to the above Remark it is clear that (1.3) is a relaxed version of (1.2). The main intuition leading to the replacement of maps with plans is that the mass initially presented at a point $x$ can be split among different $y$'s. Since (1.3) is a convex minimization problem with a convex constraint the following theorem should not be surprising.

**Theorem 1.2.** *Problem* (1.3) *admits at least one solution.*

*Proof.* The proof is a simple application of the Direct Methods in the Calculus of Variations. In fact it can be easily checked that the set (1.4) is compact with respect to the weak convergence on $\mathscr{P}(\mathbb{R}^n)$ (see [95, Lemma 4.4]) [3]. Moreover, since the function $|x - y|^2$ can be approximated by an increasing sequence of continuous and bounded functions it is immediate to check that the map

$$\gamma \mapsto \int |x - y|^2 d\gamma(x, y)$$

is lower semicontinuous with respect to the weak convergence.   □

**Remark 1.3.** Notice that the above theorem always provides a solution to problem (1.3), however it can easily happen that infimum is infinite, in this case obviously any plan is a solution. We will show in Theorem 1.13, that, however, there exists always a "locally optimal" plan between $\mu$ and $\nu$. Clearly if the second moments of $\mu$ and $\nu$ are finite:

$$\int |x|^2 d\mu(x) + \int |y|^2 d\nu(y) < +\infty,$$

the infimum is finite.

Once we have proved the existence of at least a solution of (1.3), thanks to Remark 1.1, in order to prove the existence of a solution to (1.2) we have just to understand under which assumptions an optimal plan is supported on the graph of a map $T$.

---

[2] We will say that a measure $\mu$ is concentrated on $\mu$-measurable set $A$ if $\mu(A^c) = 0$. The support of a measure, spt $\mu$, is the smallest closed set on which $\mu$ is concentrated.

[3] A sequence of probability measures $\{\mu_k\}_{k\in\mathbb{N}}$ is said to be *weakly convergent* to a probability measure $\mu$ if

$$\int \varphi d\mu_k \to \int \varphi d\mu \quad \forall \varphi \in C_b(\mathbb{R}^n),$$

where $C_b(\mathbb{R}^n)$ is the set of continuous and bounded functions.

**Example 1.4.** Let us start with a discrete example. More precisely let us assume that

$$\mu = \frac{1}{h} \sum_{i=1}^{h} \delta_{x_i}, \qquad \nu = \frac{1}{h} \sum_{j=1}^{h} \delta_{y_j}.$$

In this case a transport plan $\gamma \in \Gamma(\mu, \nu)$ can be written as

$$\gamma = \sum_{i,j=1}^{h} \gamma_{ij} \delta_{(x_i, y_j)}$$

where $(\gamma_{ij})_{i,j=1\dots h}$ is a matrix whose entries $\gamma_{ij} \geq 0$ describe which is the amount of mass moved from $x_i$ to $y_j$. The condition that all a the mass is properly transferred reads as

$$\sum_{j=1}^{h} \gamma_{ij} = 1 = \sum_{i=1}^{h} \gamma_{ij}, \tag{1.5}$$

and problem (1.3) becomes

$$\min \left\{ \sum_{i,j} |x_i - y_j|^2 \gamma_{ij} : \quad \gamma_{ij} \geq 0 \text{ satisfy (1.5)} \right\}.$$

Since the above is a finite dimensional linear optimization problem with a convex constraint, it is easy to see that the minimum is achieved in one of the extreme point of the convex set composed by the matrices with positive entries which satisfy the constraint (1.5). Birkhoff's Theorem says that a point is extremal if and only if it is represented by a permutation matrix, *i.e.* a matrix whose entries are just 0 or 1. These clearly correspond to transport maps.

Let us now investigate under which assumptions such maps are optimal. Let us assume that $y_i = T(x_i)$, $T$ is optimal. Since any other map can be obtained by $T$ simply rearranging the $y_i$ the optimality condition read as

$$\sum_{i=1}^{h} |x_i - y_i|^2 \leq \sum_{i=1}^{h} |x_i - y_{\sigma(i)}|^2 \qquad \sigma \in \mathcal{S}_h, \tag{1.6}$$

where $\mathcal{S}_h$ is the set of permutations of $h$ objects. Moreover, by elementary computations, the above condition is equivalent to the following: for every $k \leq h$ and for all choices of distinct $i_1, \dots, i_k$ in $\{1, \dots, h\}$,

$$\sum_{m=1}^{k} |x_{i_m} - y_{i_m}|^2 \leq \sum_{m=1}^{k} |x_{i_m} - y_{i_{m+1}}|^2, \tag{1.7}$$

where $i_{k+1} = i_1$. We have hence found a necessary and sufficient condition for the optimality in the discrete setting.

The above example suggests the following definition

**Definition 1.5.** A set $\Gamma \subset \mathbb{R}^n \times \mathbb{R}^n$ is said *cyclically monotone* if for every $m \geq 1$ and for all $(x_1, y_1), \ldots, (x_m, y_m))$ in $\Gamma$ it holds

$$\sum_{i=1}^{m} |x_i - y_i|^2 \leq \sum_{i=1}^{m} |x_i - y_{i+1}|^2 \qquad (1.8)$$

where $i_{m+1} := i_1$.

Expanding the squares, (1.8) is equivalent to

$$\sum_{i=1}^{m} y_i \cdot (x_{i+1} - x_i) \leq 0 \qquad (1.9)$$

with the usual convention $i_{m+1} = i_1$. We will see later that even in the continuous case, under some mild assumption, a plan $\gamma$ is optimal if and only if it is concentrated on a cyclically monotone set. Thus in order to solve (1.2) we need to understand cyclically monotone sets. The following Theorem, due to Rockafellar, characterizes such sets. We recall that the subdifferential of a proper lower semicontinuous convex function is given by

$$\partial u(x) = \left\{ y \in \mathbb{R}^n : \quad u(z) \geq u(x) + y \cdot (z - x) \text{ for all } z \in \mathbb{R}^n \right\},$$

see Appendix A. With a slight abuse of notation we will write $\partial u$ to denote its graph:

$$\partial u = \left\{ (x, y) : \quad y \in \partial u(x) \right\} \subset \mathbb{R}^n \times \mathbb{R}^n.$$

**Theorem 1.6 (Rockafellar).** *A set $\Gamma \subset \mathbb{R}^n \times \mathbb{R}^n$ is cyclically monotone if and only if it is included in the graph of the subdifferential of some proper convex and lower semicontinuous function $u$, in symbols $\Gamma \subset \partial u$.*

*Proof.* Let $u$ be a convex and lower semicontinuous function and let $(x_i, y_i) \in \partial u$, by definition of subdifferential for all $m \in \mathbb{N}$ and for all $i = 1, \ldots, m$

$$u(x_{i+1}) \geq u(x_i) + y_i \cdot (x_{i+1} - x_i)$$

with $i_{m+1} = i_1$. Summing the above relations we obtain (1.9), thus $\partial u$ is a cyclically monotone set and so is any of its subsets.

To prove the converse let us pick $(x_0, y_0) \in \Gamma$ and define

$$u(x) := \sup \left\{ y_m \cdot (x - x_m) + \cdots + y_0 \cdot (x_1 - x_0) : \quad (x_1, y_1), \ldots, (x_m, y_m) \in \Gamma \right\}.$$

Being the supremum of affine functions $u$ is clearly convex and lower semicontinuous. To see that $u$ is proper notice that, choosing $m = 1$ and $(x_1, y_1) = (x_0, y_0)$ in the definition of $u$, $u(x_0) \geq 0$. By cyclical monotonicity $u(x_0) \leq 0$ and thus $u$ is proper. To see that $\Gamma$ is included in $\partial u$ we have to show that, if $(x, y) \in \Gamma$,

$$u(z) \geq u(x) + y \cdot (z - x) \quad \forall z \in \mathbb{R}^n.$$

To see this notice that for all $\varepsilon > 0$ there exist $m \in \mathbb{N}$ and $(x_1, y_1), \ldots$ $\ldots, (x_m, y_m) \in \Gamma$ such that

$$u(x)+y\cdot(z-x) \leq y_m\cdot(x-x_m)+\cdots+y_0\cdot(x_1-x_0)+y\cdot(z-x)+\varepsilon \leq u(z)+\varepsilon,$$

where, in the last inequality, we have exploited the fact that $(x, y) \in \Gamma$ and the definition of $u$. Being $\varepsilon$ arbitrary this concludes the proof. $\qquad\square$

We now sketch the proof of the following theorem

**Theorem 1.7.** *Let* $\mu, \nu$ *be two probability measures with finite second moments, i.e.*

$$\int |x|^2 d\mu(x) + \int |y|^2 d\nu(y) < \infty. \tag{1.10}$$

*Then a plan* $\gamma \in \Gamma(\mu, \nu)$ *is optimal in* (1.3) *if and only if* spt $\gamma$ *is cyclically monotone.*

*Proof.*

*Step* 1 : *Necessity.* Let us assume that $\gamma$ is optimal and that there exist points $(x_1, y_1), \ldots, (x_m, y_m)$ in spt $\gamma$ such that

$$\sum_{i=1}^m |x_i - y_i|^2 > \sum_{i=1}^m |x_i - y_{i+1}|^2.$$

The idea is that moving mass from $x_i$ to $y_{i+1}$ instead of to $y_i$ is more convenient. To formalize this, let $U_i$, $V_i$ be neighborhoods of $x_i$ and $y_i$ such that

$$m_i := \gamma(U_i \times V_i) > 0$$

and

$$\sum_{i=1}^m |u_i - v_i|^2 > \sum_{i=1}^m |u_i - v_{i+1}|^2 \quad \forall u_i \in U_i, \ v_i \in V_i.$$

Let us consider the probability space $(\Omega, \mathbf{P})$ where

$$\Omega := \prod_i U_i \times V_i \quad \text{and} \quad \mathbf{P} := \prod_i \frac{\gamma \llcorner U_i \times V_i}{m_i}$$

and let, with a slightly abuse of notation, $u_i$ and $v_i$ be the coordinate maps from $\Omega$ to, respectively, $U_i$ and $V_i$. If we define the new plan

$$\tilde{\gamma} := \gamma + \frac{\min m_i}{m} \sum_{i=1}^{m} \left( (u_i, v_{i+1})_\sharp \mathbf{P} - (u_i, v_i)_\sharp \mathbf{P} \right),$$

a direct computation (see [11, Theorem 1.13]) shows that $\tilde{\gamma}$ is admissible and that

$$\int |x - y|^2 d\,\tilde{\gamma} < \int |x - y|^2 d\,\gamma,$$

in contradiction with the optimality of $\gamma$.

*Step* 2 : *Sufficiency* Let us assume that spt $\gamma$ is cyclically monotone. By Theorem 1.6, there exists a proper convex and lower semicontinuous function $u$ such that

$$\text{spt}\,\gamma \subset \partial u.$$

Let $u^*$ be the convex conjugate of $u$, thus

$$u(x) + u^*(y) \geq x \cdot y \tag{1.11}$$

with equality on $\partial u \supset \text{spt}\,\gamma$. Moreover since $u$ and $u^*$ are proper, condition (1.10) implies that, for any $\tilde{\gamma} \in \Gamma(\mu, \nu)$,

$$\min\{u, 0\}, \quad \min\{u^*, 0\} \in L^1(\tilde{\gamma}).$$

Thus, integrating (1.11) and using the above relations to split the integral, we obtain

$$\begin{aligned} \int x \cdot y \, d\tilde{\gamma} &\leq \int (u(x) + u^*(y)) \, d\tilde{\gamma} \\ &= \int u(x) \, d\tilde{\gamma} + \int u^*(y) \, d\tilde{\gamma} \\ &= \int u(x) \, d\mu + \int u^*(y) \, d\nu \\ &= \int u(x) \, d\gamma + \int u^*(y) \, d\gamma \\ &= \int (u(x) + u^*(y)) \, d\gamma \\ &= \int x \cdot y \, d\gamma. \end{aligned}$$

Using the finiteness of the second moments of $\mu$ and $\nu$ it is immediate to see that, adding the squares, the above inequality implies the optimality of $\gamma$. $\qquad\square$

We are now ready to give a proof of the following Theorem, due to Brenier [16].

**Theorem 1.8 (Brenier).** *Assume that $\mu$ and $\nu$ satisfy (1.10). Suppose that $\mu$ is absolutely continuous with respect to the Lebesgue measure, then there exists a unique plan $\gamma$ solution to (1.3). Moreover the plan $\gamma$ is induced by the gradient of a convex function $u$, that is $\gamma = (\mathrm{Id} \times \nabla u)_{\sharp}\mu$ and thus $\nabla u$ is also a solution to (1.2). Assume moreover that also $\nu$ is absolutely continuous with respect to the Lebesgue measure, then $\gamma = (\nabla u^* \times \mathrm{Id})_{\sharp}\nu$. In particular for $\mu$-almost every $x$ and $\nu$-almost every $y$, respectively,*

$$\nabla u^*\big(\nabla u(x)\big) = x, \qquad \nabla u\big(\nabla u^*(y)\big) = y.$$

*Proof.* Thanks to Theorem 1.2 we know that there exists an optimal plan $\gamma$, by Theorem 1.7 we now that spt $\gamma$ is cyclically monotone and by Rockafellar Theorem we deduce that spt $\gamma \subset \partial u$ for some convex proper and lower semicontinuous function. Since, see Appendix A,

$$\partial u \subset \{u < +\infty\} \times \mathbb{R}^n,$$

and $\partial\{u < +\infty\}$ is Lebesgue negligible,

$$\mu\big(\{u < +\infty\}\big) = 1.$$

Let

$$E = \{\text{Points of non differentiability of } u\},$$

since $E$ is Lebesgue negligible, by our assumptions on $\mu$

$$0 = \mu(E) = \gamma(E \times \mathbb{R}^n).$$

In conclusion $\gamma$ is concentrated on

$$\partial u \setminus (E \times \mathbb{R}^n) = \text{Graph of } \nabla u.$$

By Remark 1.1 we deduce that $\gamma = (\mathrm{Id} \times \nabla u)_{\sharp}\mu$. This proves the existence part of the claim. To obtain the uniqueness just notice that the above reasoning gives that every optimal plan is concentrated on the graph of a map. Assuming the existence of two optimal maps

$$\gamma_1 = (\mathrm{Id} \times T_1)_{\sharp}\mu, \qquad \gamma_2 = (\mathrm{Id} \times T_2)_{\sharp}\mu,$$

we see that

$$\bar{\gamma} := \frac{1}{2}\gamma_1 + \frac{1}{2}\gamma_2$$

is still optimal but not concentrated on a graph, unless $T_1 = T_2$ $\mu$-a.e.

To prove the second part of the claim just notice that, as sets, $\partial u = \partial u^*$ and that, if a plan $\gamma$ is optimal between $\mu$ and $\nu$, its "inversion",

$$R_{\sharp}\gamma, \quad \text{where } R(x, y) = (y, x),$$

is optimal between $\nu$ and $\mu$. □

**Remark 1.9.** Notice that the above proof shows that $\nabla u$ is uniquely determined $\mu$ almost everywhere. This is obviously the best uniqueness one can hope for, as simple examples show.

**Remark 1.10.** It is clear from the proof of the above theorem that the right condition on $\mu$ is that $\mu$ does not charge the set of non differentiability points of convex functions. Since it can be shown that such set is $(n - 1)$-rectifiable[4], it would have been enough to ask that $\mu$ does not charge rectifiable sets (see [11, 95]).

We can now prove Brenier Polar Factorization Theorem, a Lagrangian version of the Helmoltz Decomposition Theorem, see the remark at the end of the proof.

**Theorem 1.11.** *Let $\Omega$ be a bounded open set of $\mathbb{R}^n$ and $\mathscr{L}_{\Omega}^n$ be the normalized Lebesgue measure on $\Omega$. Let $S \in L^2(\Omega; \mathbb{R}^n)$ be such that $S_{\sharp}\mathscr{L}_{\Omega}^n$ is absolutely continuous with respect to the Lebesgue measure. Then there exist a unique gradient of convex function $\nabla u$ and a unique measure preserving map $s$ of $\Omega$ into itself[5] such that $S = (\nabla u) \circ s$. Moreover $s$ is the $L^2$ projection of $S$ on the set of measure preserving maps $S(\Omega)$,*

$$\int |S - s|^2 \, d\mathscr{L}_{\Omega}^n = \min_{\tilde{s} \in S(\Omega)} \int |S - \tilde{s}|^2 \, d\mathscr{L}_{\Omega}^n. \tag{1.12}$$

*Proof.* Let $\nu = S_{\sharp}\mathscr{L}_{\Omega}^n$ and let $\nabla u^*$ be the optimal map from $\nu$ to $\mathscr{L}_{\Omega}^n$ whose existence is given by Theorem 1.8. If we define $s := (\nabla u^*) \circ S$ we immediately see that $s \in S(\Omega)$ and that $S = (\nabla u) \circ s$. To show (1.12) we claim that

$$\inf_{\tilde{s} \in S(\Omega)} \int |S - \tilde{s}|^2 \, d\mathscr{L}_{\Omega}^n = \inf_{\gamma \in \Gamma(\mathscr{L}_{\Omega}^n, \nu)} \int |x - y|^2 \, d\gamma.$$

---

[4] Recall that a set $M$ is said $(n - 1)$-rectifiable if there exists a countable family of $C^1$ manifolds $M_i$ such that

$$\mathcal{H}^{n-1}\left(M \setminus \bigcup_i M_i\right) = 0.$$

[5] A map is said to be (Lebesgue) measure preserving if

$$s_{\sharp}\mathscr{L}_{\Omega}^n = \mathscr{L}_{\Omega}^n.$$

The set of measure preserving transformation of $\Omega$ will be denoted with $S(\Omega)$.

Since, for every $\tilde{s} \in S(\Omega)$, $(\tilde{s}, S)_{\sharp} \mathscr{L}^n_\Omega \in \Gamma(\mathscr{L}^n_\Omega, \nu)$ we clearly have that the first infimum is bigger or equal than the second one. The reverse inequality follows from the above factorization since, by Theorem 1.8,

$$\inf_{\gamma \in \Gamma(\mathscr{L}^n_\Omega, \nu)} \int |x - y|^2 \, d\gamma = \int |\nabla u^*(y) - y|^2 \, dS_{\sharp} \mathscr{L}^n_\Omega = \int |s - S|^2 \, d\mathscr{L}^n_\Omega.$$

To prove the uniqueness part notice that if $S = (\nabla \bar{u}) \circ \bar{s}$, then $\nabla \bar{u}$ is optimal between $\mathscr{L}^n_\Omega$ and $\nu$ and thus, by the uniqueness part of Theorem 1.8, it has to coincide with $\nabla u$ $\mathscr{L}^n_\Omega$ almost everywhere. $\square$

**Remark 1.12.** Let us assume that $S$ is a perturbation of the identity, *i.e.*

$$S = \mathrm{Id} + \varepsilon w + o(\varepsilon).$$

Then both $s$ and $\nabla u$ are perturbations of the identity

$$\nabla u = \mathrm{Id} + \varepsilon \nabla \psi + o(\varepsilon),$$
$$s = \mathrm{Id} + \varepsilon v + o(\varepsilon).$$

Condition $s \in S(\Omega)$ implies that $v$ is divergence-free and the polar factorization becomes

$$w = \nabla \psi + v \qquad \nabla \cdot v = 0,$$

which is the classical Helmoltz decomposition of a vector field.

Up to now we have only considered probability measures defined on $\mathbb{R}^n$ with finite second moments. In the applications to the semigeostrophic equations that we will consider in Chapter 5 we will also need the following theorem, due to McCann, dealing with the case of measures without moments. It says that, in any case, if the source measure is absolutely continuous with respect to the Lebesgue measure there exists an (essentially unique) "optimal" maps, *i.e.* the gradient of a convex function such that $(\nabla u)_{\sharp} \mu = \nu$.

**Theorem 1.13 (McCann [82]).** *Let $\mu$, $\nu \in \mathscr{P}(\mathbb{R}^n)$, assume that $\mu$ is absolutely continuous with respect to $\mathscr{L}^n$. Then there exists a unique (in the sense of Remark 1.9) gradient of a convex function $\nabla u$ such that $(\nabla u)_{\sharp} \mu = \nu$. If in addition $\nu$ is absolutely continuous with respect to $\mathscr{L}^n$ it also holds $(\nabla u^*)_{\sharp} \nu = \mu$.*

*Proof.* We only sketch the proof. Following the reasoning of the proof of Theorem 1.8 to show the existence part we only have to prove the existence of an admissible plan $\gamma \in \Gamma(\mu, \nu)$ whose support is cyclically monotone. This is done by approximating $\mu$ and $\nu$ by a sequence of discrete measures

$$\mu_k = \frac{1}{k}\sum_{i=1}^{k}\delta_{x_i} \quad \nu_k = \frac{1}{k}\sum_{i=1}^{k}\delta_{y_i}.$$

By example 1.4 there exists a cyclically monotone admissible plan $\gamma_k \in \Gamma(\mu_k, \nu_k)$. An easy computation shows that any weak cluster point of the sequence $\{\gamma_k\}_{k\in\mathbb{N}}$ is a cyclically monotone transference plan between $\mu$ and $\nu$. The proof of the uniqueness is more subtle and we refer to [82]. □

We close this section with the following stability theorem about optimal plans and optimal maps. Its easy proof is based on the cyclical monotoniticy of the support of the optimal plan (already used in the proof of Theorem 1.13). A more refined stability theorem will be proved in Chapter 4.

**Theorem 1.14.** *Let $\{\mu_k\}_{k\in\mathbb{N}}$ and $\{\nu_k\}_{k\in\mathbb{N}}$ be sequences of probability measures converging, respectively, to $\mu$ and $\nu$ and let $\gamma_k$ be cyclically monotone transference plans from $\mu_k$ to $nu_k$. Then any weak cluster point of the sequence $\{\gamma_k\}_{k\in\mathbb{N}}$ is a cyclically monotone transference plan between $\mu$ and $\nu$. Moreover, if $\mu_k = \mu$ and is absolutely continuous with respect to $\mathscr{L}^n$, the sequence of optimal maps $T_k$ between $\mu$ and $\nu_k$ converges in $\mu$-measure to the optimal map $T$ between $\mu$ and $\nu$, i.e. for all $\varepsilon > 0$*

$$\lim_{k\to\infty}\mu\Big(\big\{x : \quad |T_k(x) - T(x)| \ge \varepsilon\big\}\Big) = 0.$$

*Proof.* The first part of the statement is immediate and it is based on the simple observation that if $z \in \operatorname{spt}\gamma$ there exists a sequence of points $z_k \in \operatorname{spt}\gamma_k$ converging to $z$. The proof of the second part follows from the general statement that if

$$(\operatorname{Id} \times T_k)_{\sharp}\mu \rightharpoonup (\operatorname{Id} \times T)_{\sharp}\mu,$$

then $T_k$ converges in $\mu$-measure to $T$. To see this recall that by Lusin Theorem, for very $\delta > 0$ there exists a compact set $K$ such that $\mu(\mathbb{R}^n \setminus K) \le \delta$ and $T$ restricted to $K$ is continuous. Let us consider the upper semicontinuous and bounded function [6]

$$\varphi(x, y) = \mathbf{1}_K(x)\min\big\{1, |y - T(x)|/\varepsilon\big\}.$$

---

[6] With $\mathbf{1}_A$ we denote the characteristic function of a set $A$.

By approximating $\varphi$ with a decreasing sequence of continuous and bounded function we see that

$$0 = \int \varphi(x, y)d\, (\mathrm{Id} \times T)_\sharp \mu \geq \limsup_{k \to \infty} \int \varphi(x, y)d\, (\mathrm{Id} \times T_k)_\sharp \mu.$$

Hence

$$\limsup_{k \to \infty} \mu\Big(\big\{x : \quad |T_k(x) - T(x)| \geq \varepsilon\big\}\Big)$$

$$\leq \limsup_{k \to \infty} \mu\Big(\big\{x \in K : \quad |T_k(x) - T(x)| \geq \varepsilon\big\}\Big) + \delta$$

$$\leq \limsup_{k \to \infty} \int \varphi(x, y)d\, (\mathrm{Id} \times T_k)_\sharp \mu + \delta = \delta.$$

The conclusion follows letting $\delta \to 0$.   □

## 1.2. Brenier vs Aleksandrov solutions to the Monge-Ampère equation

### 1.2.1. Brenier solutions

In this section we start investigating the regularity of optimal transport maps. We assume that both $\mu$ and $\nu$ are absolutely continuous with respect to the Lebesgue measure,

$$\mu = \rho_1 \mathscr{L}^n, \qquad \nu = \rho_2 \mathscr{L}^n \tag{1.13}$$

and that

$$\mathrm{spt}\, \rho_1 = \overline{\Omega}_1, \qquad \mathrm{spt}\, \rho_2 = \overline{\Omega}_2$$

with $\Omega_1$ and $\Omega_2$ open and bounded subsets of $\mathbb{R}^n$ with $\mathscr{L}^n(\partial\Omega_1) = \mathscr{L}^n(\partial\Omega_2) = 0$. We will also assume that, on their support,

$$\lambda \leq \rho_1, \rho_2 \leq 1/\lambda \tag{1.14}$$

for some positive constant $\lambda$.

Under this assumptions Theorem 1.8 ensures the existence of an optimal map between $\mu$ and $\nu$, moreover the map is given by the gradient of a convex function $u$. It is clear that any regularity of $u$ immediately translates in regularity of $T$. We will hence start the study of the regularity of $u$. The first point we would like to show is how the condition

$$(\nabla u)_\sharp \mu = \nu$$

implies that $u$ satisfies a suitable Monge-Ampère type equation. To see this we first recall the following version of the Area Formula (see [49, Corollary 3.2.20]).

**Theorem 1.15 (Area Formula).** *Let $T$ be differentiable almost everywhere. Let $\Sigma = \mathrm{Dom}(\nabla T)$, then for every Borel and bounded function $\varphi$*

$$\int_{\Sigma} \varphi(x)|\det \nabla T|(x)\, dx = \int_{\mathbb{R}^n} \left( \sum_{x \in \Sigma \cap T^{-1}(y)} \varphi(x) \right) dy. \qquad (1.15)$$

Since $u$ is convex, Aleksandrov Theorem (see Theorem A.5 ) implies that $\nabla u$ is twice differentiable almost everywhere on $\mathrm{Dom}(\nabla u)$, thus we an apply the above Theorem to $T = \nabla u$ restricted to $\mathrm{Dom}(\nabla u)$. Taking into account (1.13) and the relation $(\nabla u)_{\sharp}\mu = \nu$, we infer that for all bounded and Borel functions $\varphi$

$$\int \varphi(\nabla u(x))\rho_1(x)\, dx = \int \varphi(y)\rho_2(y)\, dy.$$

Since $\nabla u \circ \nabla u^* = \mathrm{Id}$ almost everywhere on $\Omega_2$,

$$\left| \left\{ y \in \Omega_2 : \#\{(\nabla u)^{-1}(y)\} \geq 2 \right\} \right| = 0. \quad (^7)$$

We can thus apply Theorem 1.15 to deduce

$$\int \varphi(\nabla u(x))\rho_1(x)\, dx = \int \varphi(\nabla u(x))\rho_2(\nabla u(x)) \det \nabla^2 u(x)\, dx.$$

Applying the above relation to $\varphi = \tilde{\varphi} \circ \nabla u^*$ and recalling that, $\nabla u^* \circ \nabla u = \mathrm{Id}$ almost everywhere in $\Omega_1$, we obtain

$$\int \tilde{\varphi}(x)\rho_1(x)dx = \int \tilde{\varphi}(x)\rho_2(\nabla u(x)) \det \nabla^2 u(x)dx \quad \forall \tilde{\varphi} \text{ Borel and bounded.}$$

In conclusion it holds

$$\det \nabla^2 u = \frac{\rho_1}{\rho_2 \circ \nabla u} \quad \text{a.e. in } \Omega_1. \qquad (1.16)$$

The above equation shall be considered together with the following "boundary condition"

$$\nabla u(\Omega_1) \subset \Omega_2. \qquad (1.17)$$

We are going to call a convex function satisfying (1.16) and (1.17) a *Brenier solution* to the Monge-Ampère equation.

We will now show that a Brenier solution to the Monge-Ampère equation, in general, can be disccontinuos even if the densities $f$ and $g$ are smooth on their support.

---

[7] In the sequel we are going to use the notation $|A|$ for the outer Lebesgue measure of $A$.

**Example 1.16.** Let us consider

$$\rho_1 = \mathbf{1}_B/|B| \quad \rho_2 = (\mathbf{1}_{B^+} + \mathbf{1}_{B^-})/|B|,$$

where $B$ is the unitary ball $B = \{|x| < 1\}$ and $B^\pm$ are two shifted half balls

$$B^+ = e_1 + B \cap \{x_1 > 0\}, \qquad B^- = -e_1 + B \cap \{x_1 < 0\}.$$

Clearly $\rho_1$ and $\rho_2$ are smooth on their supports, but from obviously topological reason there cannot be a continuous map pushing forward $\rho_1 \mathscr{L}^n$ to $\rho_2 \mathscr{L}^n$. The optimal map $T$ can be also explicitly computed thanks to the necessary and sufficient optimality conditions, $T = \nabla u$ where $u(x) = |x|^2/2 + |x_1|$, which is clearly discontinuous (see Figure 1.1).

$$T = \nabla \left( \frac{|x|^2}{2} + |x_1| \right)$$

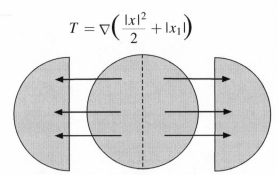

**Figure 1.1.** A discontinuous optimal map.

One might think that the obstruction to the regularity of the optimal map in the above example is given by the lack of connectness of the support of the target measure. Actually one can modify the above example considering as target measure the normalized Lebesgue measure restricted to the set $D_\varepsilon$, where $D_\varepsilon$ is the set obtained joining $B^+$ and $B^-$ with a strip of width $\varepsilon$. By Theorem 1.14 we deduce that the optimal maps $T_\varepsilon$ converge in measure to the optimal map of the above example. Using this, one can easily show that the continuity of $T_\varepsilon$ would contradict the cyclical monotoniticy of its graph (see [95, Theorem 12.3] for more details). As we will show in the sequel the right assumption on the support of the target measure is *convexity*.

### 1.2.2. Aleksandrov solutions

We will now introduce the concept of *Aleksandrov solution* of the Monge-Ampère equation. The study of their properties will be the topic of the next chapters, here we simply recall the definition and show under which assumption a Brenier solution is an Aleksandrov solution. A good reference for Aleksandorv solutions of the Monge-Ampère equation is the book of Gutierrez, [67].

First we recall that the subdifferential of a (finite) convex function $u$ defined on a convex open domain $\Omega$ is defined as:

$$\partial u(x) = \{p \in \mathbb{R}^n : \quad u(y) \geq u(x) + p \cdot (y - x) \quad \forall y \in \Omega\}.$$

We define the *Monge-Ampère measure* of $u$ in the following way: for every set $E \subset \Omega$,

$$\mu_u(E) = |\partial u(E)| = \left| \bigcup_{x \in E} \partial u(x) \right|, \tag{1.18}$$

here $| \cdot |$ is the Lebesgue *outer* measure. Notice that in case $u \in C^2(\Omega)$, the Area Formula implies

$$\mu_u = \det \nabla^2 u(x) \mathscr{L}^n.$$

We now show that the restriction of $\mu_u$ to the Borel $\sigma$-algebra is actually a measure and that its absolutely continuous part with respect to the Lebesgue measure is given by $\det \nabla^2 u(x)\, dx$. We start with the following simple lemma.

**Lemma 1.17.** *Let $u : \Omega \to \mathbb{R}$ be a continuous convex function defined on an open convex set $\Omega$. Let us consider the set*

$$S = \{p \in \mathbb{R}^n : \text{there exist } x \text{ and } y \text{ in } \Omega, \ x \neq y, \text{ such that } p \in \partial u(x) \cap \partial u(y)\}.$$

*Then $|S| = 0$.*

*Proof.* Let us first assume that $\Omega$ is bounded and consider the convex conjugate

$$u^*(p) = \sup_{x \in \Omega} \left(x \cdot p - u(x)\right).$$

Then $u^*$ is finite everywhere and $p \in \partial u(x)$ if and only if $x \in \partial u^*(p)$ (see Appendix A). Thus

$$S \subset \{p : \text{ s.t. } \#(\partial u^*(p)) > 1\} = \{\text{Points of non differentiability of } u^*\}.$$

Being $u^*$ a (finite) convex function it is locally Lipschitz and thus differentiable almost everywhere, hence $|S| = 0$. In case $\Omega$ is unbounded, we can write it as an increasing union of convex and bounded set $\Omega_k$ and define

$$S_k = \Big\{ p \in \mathbb{R}^n : \text{there exist } x \text{ and } y \text{ in } \Omega_k, x \neq y \text{ such that}$$

$$p \in \partial(u \llcorner \Omega_k)(x) \cap \partial(u \llcorner \Omega_k)(y) \Big\}.$$

Since $S \subset \cup_k S_k$ and, by the first part of the proof, $|S_k| = 0$, we conclude the proof. □

**Lemma 1.18.** *Let Let* $u : \Omega \to \mathbb{R}$ *be a continuous convex function defined on an open convex set* $\Omega$. *Let us consider*

$$\mathcal{M} = \big\{ E \subset \Omega \text{ such that } \partial u(E) \text{ is Lebesgue measurable.} \big\},$$

*then* $\mathcal{M}$ *is a* $\sigma$-*algebra containing all the Borel subsets of* $\Omega$. *Moreover* $\mu_u : \mathcal{M} \to [0, +\infty]$ *is a measure and the density of its absolutely continuous part with respect to* $\mathcal{L}^n$ *is given by*

$$\frac{d\,\mu_u}{d\,\mathcal{L}^n} = \det \nabla^2 u. \tag{1.19}$$

*Proof.* First notice that if $K \subset \Omega$ is compact, then $\partial u(K)$ is compact. Indeed the closureof $\partial u(K)$ follows from the closure of the subdifferential [8] while the boundedness follows from the following standard estimates for convex functions (see Appendix A):

$$\sup_{p \in \partial u(K)} |p| \leq \operatorname*{osc}_{K_\delta} u/\delta,$$

where $K_\delta = \{x : \operatorname{dist}(x, K) \leq \delta\} \subset \Omega$ for small $\delta$. Thus in order to prove the first part of the claim we only have to show that $\mathcal{M}$ is a $\sigma$-algebra. Clearly $\partial u(\cup E_k) = \cup \partial u(E_k)$ and thus $\mathcal{M}$ is closed with respect to countable union, writing $\Omega$ as countable union of compact sets we also see that $\Omega \in \mathcal{M}$. We need only to show that if $E \in \mathcal{M}$, then $\Omega \setminus E \in \mathcal{M}$. Let us write

$$\partial u(\Omega \setminus E) = \big(\partial u(\Omega) \setminus \partial u(E)\big) \cup \big(\partial u(\Omega \setminus E) \cap \partial u(E)\big).$$

By Lemma 1.17 we see that the last set is Lebesgue negligible (and hence Lebesgue measurable), thus $\Omega \setminus E \in \mathcal{M}$.

---

[8] Which means that if $p_k \in \partial u(x_k)$, $x_k \to x$ and $p_k \to p$ then $p \in \partial u(x)$

Since $\mu_u$ is an outer measure to show it is a measure it is enough to prove that it is finitely additive on $\mathcal{M}$. Let $E$ and $F$ be disjoint sets in $\mathcal{M}$, then

$$\partial u(E \cup F) = \big(\partial u(E) \setminus (\partial u(F) \cap \partial u(E))\big) \cup \big(\partial u(F) \setminus (\partial u(F) \cap \partial u(E))\big)$$
$$\cup \big(\partial u(F) \cap \partial u(E)\big),$$

where the union is disjoint. Since $E \cap F = \emptyset$, Lemma 1.17 implies $|\partial u(F) \cap \partial u(E)| = 0$ from which finite additivity follows.

We now want to show (1.19). Let us consider the set

$$\mathcal{S} = \{x \in \Omega \text{ such that } \nabla u(x) \text{ and } \nabla^2 u(x) \text{ exist}\}.$$

Clearly $\mathcal{S}$ is of full (Lebesgue) measure in $\Omega$ and

$$\mu_u^a = \mu_u^a \llcorner \mathcal{S}. \tag{1.20}$$

If $E \subset \mathcal{S}$ is Borel measurable, the Area Formula (1.15) implies

$$\int_E \det \nabla^2 u(x)\,dx = \int_{\mathbb{R}^n} \#\{x \in E : \nabla u(x) = y\}\,dy.$$

By Lemma 1.17, the set of $y$ such that $\#\{x \in E : \nabla u(x) = y\} > 1$ has zero measure and thus

$$\int_E \det \nabla^2 u(x)\,dx = \int_{\mathbb{R}^n} \mathbf{1}_{\nabla u(E)}\,dy = |\nabla u(E)| = |\partial u(E)| = \mu_u(E).$$

Thus $\mu_u \llcorner \mathcal{S} = \det \nabla^2 u\,dx$ which, together with (1.20), proves (1.19). $\qquad \square$

We are now ready to give the following definition.

**Definition 1.19.** Given an open convex set $\Omega$ and a Borel measure $\mu$ on $\Omega$, a convex and continuous function $u : \Omega \to \mathbb{R}$ is said an *Aleksandrov solution* to the Monge-Ampère equation

$$\det D^2 u = \mu,$$

if $\mu = \mu_u$ as Borel measures.

**Remark 1.20.** In view of the above definition, in the sequel we will use both the notations $\mu_u$ and $\det D^2 u$ to denote the Monge-Ampère measure associated to $u$. We will instead use the notation $\det \nabla^2 u$ *only* to denote the determinant of $\nabla^2 u$, the density of the absolutely continuous part of the distributional Hessian of $u$, $D^2 u$ (see Appendix A).

**Example 1.21.** Let $u(x) = |x|$, then

$$\partial u(x) = \begin{cases} \{x/|x|\} & \text{if } x \neq 0 \\ \{p : |p| \leq 1\} & \text{if } x = 0. \end{cases}$$

Thus $\mu_u = |B_1|\delta_0$.

**Example 1.22.** Let $u(x) = |x|^2/2 + |x_1|$, then (writing $x = (x_1, x')$)

$$\partial u(x) = \begin{cases} \{x + e_1\} & \text{if } x_1 > 0 \\ \{x - e_1\} & \text{if } x_1 < 0 \\ \{(t, x') : |t| \leq 1\} & \text{if } x_1 = 0. \end{cases}$$

Thus $\mu_u = \mathscr{L}^n + \mathcal{H}^{n-1} \llcorner \{x_1 = 0\}$.

The above example, in combination with example (1.16), shows that the map $u(x) = |x|^2/2 + |x_1|$ is a Brenier solution to the Monge-Ampère equation

$$\det \nabla^2 u = \mathbf{1}_B/(\mathbf{1}_{B^+} \circ \nabla u + \mathbf{1}_{B^-} \circ \nabla u)$$

but not an Aleksandrov solution. The reason for this gap is the following. The singular part of the Monge-Ampère measure associated to $u$ is due to the image through the subdifferential map of the line $\{x_1 = 0\}$. Since this image is not contained in the support of $\rho_2$, the relation $(\nabla u)_\sharp(\rho_1 \mathscr{L}^n) = \rho_2 \mathscr{L}^n$ cannot give any control on this singular part. Since, as it immediately follows from the definition, $\partial u(x)$ is a convex set it is clear that convexity of the support of the target measure is the right notion to ensure a control on the singular part of $\mu_u$.

**Proposition 1.23.** *Let us assume that $\mu = \rho_1 \mathscr{L}^n$ and $\nu = \rho_2 \mathscr{L}^n$ satisfy the assumption at the beginning of the section and that $\Omega_2$ is convex. If $T = \nabla u$ is the optimal map between $\mu$ and $\nu$, then $u$ coincides on $\Omega_1$ with an Aleksandrov solution of*

$$\det D^2 u = \frac{\rho_1}{\rho_2 \circ \nabla u} \mathscr{L}^n,$$

*where we understand that $\rho_1$ is zero outside $\Omega_1$.*

*Proof.* First we define a "canonical" map $\bar{u}$ such that $(\nabla \bar{u})_\sharp \mu = \nu$ (recall Remark 1.9). To do this we define

$$\bar{u}(x) = \sup_{y \in \overline{\Omega}_2} \left( x \cdot y - u^*(y) \right). \tag{1.21}$$

Recalling that

$$u(x) = \sup_{y \in \mathbb{R}^n} \left( x \cdot y - u^*(y) \right),$$

we see that $\bar{u} \leq u$. Being $\overline{\Omega}_2$ compact and $u^*$ lower semicontinuous on this set we see that supremum in (1.21) is attained and that $\bar{u}$ is locally bounded and hence continuous. If $x \in \text{Dom}(\nabla u(x))$, then (see Appendix A)

$$u(x) = x \cdot \nabla u(x) - u^*(\nabla u(x)),$$

hence, for every $x$ such that $\nabla u(x) \in \Omega_2$, $u(x) \leq \bar{u}(x)$. Being this set dense in $\overline{\Omega}_1$ it follows that $u$ coincide with $\bar{u}$ on $\Omega_1$. In particular at any point of differentiability of $u$, $\nabla \bar{u}$ exists and equals $\nabla u$, thus $(\nabla \bar{u})_\sharp \mu = \nu$. We now show that $\bar{u}$ is, in some sense, the minimal potential satisfying the previous relation, meaning that

$$\nabla \bar{u}(\mathbb{R}^n) \subset \overline{\Omega}_2.$$

To see this let $\bar{x}$ a point of differentiability of $\bar{u}$ and let $\bar{y} \in \overline{\Omega}_2$ be a point where the supremum in (1.21) is attained, then

$$x \cdot \bar{y} - u^*(\bar{y}) \leq \bar{u}(x) = \bar{u}(\bar{x}) + \nabla \bar{u}(\bar{x}) \cdot (x - \bar{x}) + o(|x - \bar{x}|)$$
$$= \bar{x} \cdot \bar{y} - u^*(\bar{y}) + \nabla \bar{u}(\bar{x}) \cdot (x - \bar{x}) + o(|x - \bar{x}|).$$

Simplifying

$$\bar{y} \cdot (x - \bar{x}) \leq \nabla \bar{u}(\bar{x}) \cdot (x - \bar{x}) + o(|x - \bar{x}|) \quad \forall x,$$

which implies $\nabla \bar{u}(\bar{x}) = \bar{y} \in \overline{\Omega}_2$. Notice that for this computations we have neither used the hypothesis that $\Omega_2$ is convex nor the boundedness of $\Omega_1$.

We now show that $\bar{u}$ is an Aleksandrov solution to the Monge-Ampère equation. Since $\det \nabla^2 \bar{u}$ is the density of the absolutely continuous part of $\mu_{\bar{u}}$, inequality

$$\det D^2 \bar{u} \geq \frac{\rho_1}{\rho_2 \circ \nabla \bar{u}} \mathscr{L}^n$$

holds without any assumption on $\Omega_2$. To prove the equality we only need to show that

$$|\partial \bar{u}(\mathbb{R}^n)| \leq \int_{\Omega_1} \frac{\rho_1(x)}{\rho_2(\nabla \bar{u}(x))} \, dx.$$

Since $\partial \bar{u}(x)$ is convex and any of its extremal points $p$ can be approached by a sequence of points $p_k = \nabla \bar{u}(x_k)$, $x_k \to x$ (see Appendix A), we see that

$$\partial \bar{u}(\mathbb{R}^n) \subset \text{co}\left[\overline{\nabla \bar{u}(\mathbb{R}^n)}\right] \subset \overline{\Omega}_2,$$

thanks to the convexity of $\Omega_2$. From this it follows that

$$|\partial \bar{u}(\mathbb{R}^n)| \leq |\Omega_2| = \int_{\Omega_2} \frac{\rho_2(y)}{\rho_2(y)} \, dy = \int_{\Omega_1} \frac{\rho_1(x)}{\rho_2(\nabla \bar{u}(x))} \, dx,$$

where in the last equality we have used the definition of push-forward.

$\square$

**Remark 1.24.** Notice that the proof of the above proposition actually gives the following stronger statement: If $A \subset \Omega_1$ is such that $\partial u(A) \subset \Omega_2$ then

$$\det D^2 u \, \llcorner \, A = \frac{\rho_1(x)}{\rho_2(\nabla u(x))} \mathscr{L}^n \, \llcorner \, A.$$

Arguing as in the above proof we only have to show that

$$|\partial u(A)| = \int_A \frac{\rho_1(x)}{\rho_2(\nabla u(x))} \, dx.$$

To see this notice that for all $A \subset \Omega_1$,

$$A \cap \mathrm{Dom}(\nabla u) \subset (\nabla u)^{-1}(\partial u(A))$$

and

$$(\nabla u)^{-1}\big(\partial u(A) \cap \Omega_2\big) \setminus A$$
$$\subset (\nabla u)^{-1}\Big(\big\{ y \in \Omega_2 : \text{there exist } x_1, x_2, x_1 \neq x_2 \text{ such that}$$
$$y \in \partial u(x_1) \cap \partial u(x_2) \big\}\Big).$$

Since, by Lemma 1.17 and our assumptions on the densities, this latter set has measure zero and $\partial u(A) \subset \Omega_2$ we have

$$|\partial u(A)| = \int_{\partial u(A) \cap \Omega_2} \frac{\rho_2(y)}{\rho_2(y)} \, dy = \int_{(\nabla u)^{-1}(\partial u(A))} \frac{\rho_1(x)}{\rho_2(\nabla u(x))} \, dx$$
$$= \int_A \frac{\rho_1(x)}{\rho_2(\nabla u(x))} \, dx + \int_{(\nabla u)^{-1}(\partial u(A)) \setminus A} \frac{\rho_1(x)}{\rho_2(\nabla u(x))} \, dx$$
$$= \int_A \frac{\rho_1(x)}{\rho_2(\nabla u(x))} \, dx,$$

proving our claim. In particular, Proposition 1.23 can be extended to the case in which $\Omega_2$ is convex but unbounded.

We conclude the sections studying the behavior of sequences of Monge-Ampère measures $\mu_{u_k}$ under the uniform convergence of the functions $u_k$.

**Proposition 1.25.** *If $u_k : \Omega \to \mathbb{R}$ are convex functions locally uniformly converging to $u$, then*

$$\mu_{u_k} \overset{*}{\rightharpoonup} \mu_u \tag{1.22}$$

*as Radon measures (i.e. in the duality with $C_c(\Omega)$).*

*Proof.* It is well known (see for instance [47, Section 1.9]) that the weak-$*$ convergence as Radon measures is equivalent to the following two inequalities

$$\mu_u(A) \leq \liminf_{k \to \infty} \mu_{u_k}(A) \qquad \text{for all open sets } A \subset \Omega \tag{1.23}$$

$$\mu_u(K) \geq \limsup_{k \to \infty} \mu_{u_k}(K) \qquad \text{for all compact sets } K \subset \Omega. \tag{1.24}$$

The proof of (1.24) follows from the following relation

$$\partial u(K) \supset \limsup_{k \to \infty} \partial u_k(K) \quad \text{for all compact set } K.$$

To prove it let $p$ belong to $\limsup_{k \to \infty} \partial u_k(K)$, this means that there exists a (not relabeled) subsequence $u_k$ converging to $u$ and points $x_k \in K$ such that $p \in \partial u_k(x_k)$. Since, again up to subsequences, $x_k \to x \in K$ it is immediate to see that $p \in \partial u(x)$.

To prove (1.23) it is enough to show that if $K \subset A \Subset \Omega$, $K$ compact and $A$ open, then

$$|\partial u(K)| \leq \liminf_{k \to \infty} |\partial u_k(A)|. \tag{1.25}$$

Let $S$ be defined as in Lemma 1.17, and $p \in \partial u(K) \setminus S$, we want to show

$$p \in \partial u_k(A)$$

for $k$ large enough. Taking into account that $|S| = 0$, this implies (1.25). Since $p \in \partial u(K) \setminus S$ there exists a point $x \in K$ such that $p \in \partial u(x)$ and $p \notin \partial u(z)$ for all $z \neq x$, thus

$$u(y) > u(x) + p \cdot (y - x) \quad \forall y \in \Omega, \ y \neq x.$$

Being $\overline{A}$ compact

$$\min_{y \in \partial A} u(y) - u(x) - p \cdot (y - x) := 4\delta > 0.$$

Choosing $k_0$ such that $\|u - u_k\|_{L^\infty(A)} \leq \delta$ for $k \geq k_0$, we see that

$$y_k := \operatorname*{argmin}_{y \in \overline{A}} \left( u_k(y) - u(x) - p \cdot (y - x) \right) \in A \qquad \forall k \geq k_0.$$

This implies that, for $k \geq k_0$, $p \in \partial u_k(y_k)$ with $y_k \in A$, but this exactly means that $p$ belongs to $\liminf_k \partial u_k(A)$. $\qquad \square$

## 1.3. The case of a general cost $c(x, y)$

### 1.3.1. Existence of optimal maps

In this section we show how to solve problem (1.1), the strategy will be the same one of the case $c(x, y) = |x - y|^2$. Through the whole section we make the following assumptions on the cost and on $X$ and $Y$, they are far from being necessary (see [95, Chapter 10] for more refined results) nevertheless they will be sufficient for our goals (see Chapter 6).
Either $X$ and $Y$ are bounded open sets of $\mathbb{R}^n$ and

- **(C0)** The cost function $c : X \times Y \to \mathbb{R}$ is of class $C^2$ with $\|c\|_{C^2(X \times Y)} < \infty$.
- **(C1)** For any $x \in X$, the map $Y \ni y \mapsto -D_x c(x, y) \in \mathbb{R}^n$ is injective.
- **(C2)** For any $y \in Y$, the map $X \ni x \mapsto -D_y c(x, y) \in \mathbb{R}^n$ is injective.

Or $X = Y = M$, a compact Riemannian manifold, and $c(x, y) = d^2(x, y)/2$, the square of the Riemannian distance. In this case it is known (see [11, 95]) that $c$ satisfies a local version of **(C0)-(C2)** outside the *cut locus*

$$\text{cut}M = \bigcup_{x \in M} \{x\} \times \text{cut}_x M = \bigcup_{y \in M} \text{cut}_y M \times \{y\} \subset M \times M.$$

As in the previous section we introduce the relaxed problem

$$\inf_{\gamma \in \Gamma(\mu, \nu)} \int c(x, y) d\gamma(x, y). \tag{1.26}$$

The proof of the existence of a solution of (1.26) is identical to Theorem 1.2. Moreover, exactly as in (1.7), one can prove that spt $\gamma$ is *c-cyclically monotone set*. In this case $c$-cyclical monotonicity of a set $\Gamma \subset X \times Y$ means

$$\sum_{i=1}^m c(x_i, y_i) \leq \sum_{i=1}^m c(x_i, y_{\sigma(i)}) \qquad \forall m \in \mathbb{N}, \ (x_i, y_i) \in \Gamma, \ \sigma \in \mathcal{S}_h.$$

To characterize $c$-cyclically monotone sets we need to introduce $c$-convex function: a function $u : X \to \mathbb{R} \cup +\infty$ is said *c-convex* if it can be written as

$$u(x) = \sup_{y \in Y} \left\{ -c(x, y) + \lambda_y \right\}, \tag{1.27}$$

for some constants $\lambda_y \in \mathbb{R} \cup \{-\infty\}$. If $u : X \to \mathbb{R} \cup +\infty$ is a $c$-convex function as above, the *c-subdifferential* of $u$ at $x$ is the (nonempty) set

$$\partial_c u(x) := \{y \in Y : \ u(z) \geq -c(z, y) + c(x, y) + u(x) \quad \forall z \in X\}. \tag{1.28}$$

If $x_0 \in X$ and $y_0 \in \partial_c u(x_0)$, we will say that the function

$$C_{x_0, y_0}(\cdot) := -c(\cdot, y_0) + c(x_0, y_0) + u(x_0) \qquad (1.29)$$

is a *c-support* for $u$ at $x_0$. The version of Rockafellar Theorem for $c$-cyclically monotone sets is the following:

**Theorem 1.26.** *A set $\Gamma \subset X$ is c-cyclically monotone if and only if it is included in the c-subdifferential of a c-convex function.*

Finally, let us observe that if $c$ satisfies **(C0)** and both $X$ and $Y$ are bounded, then it follows immediately from (1.27) that $u$ is Lipschitz and semiconvex (*i.e.*, there exists a constant $C > 0$ such that $u + C|x|^2/2$ is convex). In particular, $c$-convex functions are twice differentiable a.e. and the *Frechet subdifferential* of $u$ at $x$:

$$\partial^- u(x) := \left\{ p \in \mathbb{R}^n : u(z) \geq u(x) + p \cdot (z - x) + o(|z - x|) \right\}.$$

is not empty and single valued almost everywhere (see Appendix A).

Let us assume that $y \in \partial_c u(x)$ and that $u$ is differentiable at $x$. Clearly (1.28) implies that the function:

$$z \mapsto u(z) + c(z, y)$$

has a minimum at $x$. Differentiating we get

$$\nabla u(x) = -D_x c(x, y). \qquad (1.30)$$

If $c$ satisfies **(C1)** this univocally determines $y$. Notice that in any case the following relation holds:

$$y \in \partial_c u(x) \implies -D_x c(x, y) \in \partial^- u(x). \qquad (1.31)$$

The above relations becomes more suggestive once we introduce the following definition:

**Definition 1.27.** If $c$ satisfies **(C0)-(C2)**, then we can define the *c-exponential map*:

for any $x \in X, \, y \in Y, \, p \in \mathbb{R}^n$, $\begin{cases} \text{c-exp}_x(p) = y & \Leftrightarrow & p = -D_x c(x, y) \\ \text{c*-exp}_y(p) = x & \Leftrightarrow & p = -D_y c(x, y) \end{cases}$

$$(1.32)$$

Using (1.32), we can rewrite (1.31) as

$$\partial_c u(x) \subset \text{c-exp}_x \left( \partial^- u(x) \right). \tag{1.33}$$

We notice here that in general, if $\partial^- u(x)$ contains more than one vector, the above inclusion can be strict. Equality in the above inclusion is related to the connectness of the $c$-subdifferential, a condition which turns out to be necessary (and almost sufficient) for regularity of optimal transport maps, see the discussion at the end of the Section. Finally we notice that if $c(x, y) = d^2(x, y)/2$, then c-exp coincides with the classical exponential in Riemannian geometry.

As we said if $c$ satisfies **(C0)** then $u$ is semi-convex and thus differentiable almost everywhere, see Appendix A. If $\mu$ is absolutely continuous with respect to the Lebesgue measure (or if $\mu$ does not charge rectifiable sets see Remark 1.10) then exactly as in the proof of Theorem (1.8) any optimal plan $\gamma$ is concentrated on the set

$$\partial_c u \setminus \{\text{Points of non differentiability of } u\} \times \mathbb{R}^n.$$

Thus the calculation which led to (1.30) can be actually performed and shows that $\gamma$ is concentrated on the graph of the map $T_u(x)$ defined through

$$-D_x c(x, T_u(x)) = \nabla u(x).$$

Using the $c$ exponential we can re-write the above relation as

$$T_u(x) := \text{c-exp}_x(\nabla u(x)). \tag{1.34}$$

We summarize the previous consideration in the following Theorem.

**Theorem 1.28.** *Let $c : X \times Y \to \mathbb{R}$ satisfy **(C0)-(C1)**. Given two absolutely continuous probability measures $\mu = f\mathcal{L}^n$ and $\nu = g\mathcal{L}^n$ supported on $X$ and $Y$ respectively, there exists a $c$-convex function $u : X \to \mathbb{R}$ such that $T_u : X \to Y$ is the unique optimal transport map sending $\mu$ onto $\nu$. If in addition $c$ satisfies **(C2)**, there exists a unique optimal transport map $T^*$ sending $\nu$ onto $\mu$ such that*

$$\int c(T^*(y), y) \, dv = \min_{(S^*)_{\sharp} \nu = \mu} \int c(S^*(y), y) \, dv.$$

*Moreover $T^*$ is given by*

$$T^*(y) = T_{u^c}(y) = \text{c*-exp}(\nabla u^c(y)),$$

*where*

$$u^c(y) = \sup_{x \in X} \{-c(x, y) - u(x)\},$$

*is $c^*$-convex with $c^*(y, x) := c(x, y)$. In addition*

$$T^* \circ T = \text{Id} \ \mu\text{-a.e.}, \qquad T \circ T^* = \text{Id} \ \nu\text{-a.e.} \tag{1.35}$$

In the particular case $c(x, y) = -x \cdot y$ (which is equivalent to the quadratic cost $|x - y|^2/2$), $c$-convex functions are convex and the above result gives back Brenier Theorem 1.8.

Although on compact Riemannian manifolds the cost function $c = d^2/2$ is not smooth everywhere, one can still prove existence of optimal maps [83, 54]

**Theorem 1.29 (McCann).** *Let $M$ be a Riemannian manifold, and $c = d^2/2$. Given two absolutely continuous probability measures $\mu = f \text{vol}_M$ and $\nu = g \text{vol}_M$ supported on $M$ [9], there exists a $c$-convex function $u$ : $X \to \mathbb{R}$ such that $T_u(x) = \exp_x(\nabla u(x))$ is the unique optimal transport map sending $\mu$ onto $\nu$.*

We conclude showing $c$-convex functions arising in optimal transport problems solve a Monge-Ampère type equation. Indeed by the Area Formula (1.15) and (1.35), $(T_u)_\sharp(f \mathcal{L}^n) = g \mathcal{L}^n$ gives

$$|\det(DT_u(x))| = \frac{f(x)}{g(T_u(x))} \qquad \text{a.e.} \tag{1.36}$$

In addition, the $c$-convexity of $u$ implies that, at every point $x$ where $u$ is twice differentiable,

$$D^2 u(x) + D_{xx}c\big(x, \text{c-exp}_x(\nabla u(x))\big) \geq 0. \tag{1.37}$$

Hence, by writing (1.34) as

$$-D_x c(x, T_u(x)) = \nabla u(x)$$

and differentiating the above relation with respect to $x$, we obtain

$$\det\Big(D^2 u(x) + D_{xx}c\big(x, \text{c-exp}_x(\nabla u(x))\big)\Big)$$
$$= \Big|\det\big(D_{xy}c\big(x, \text{c-exp}_x(\nabla u(x))\big)\big)\Big| \frac{f(x)}{g(\text{c-exp}_x(\nabla u(x)))} \tag{1.38}$$

at every point $x$ where $u$ it is twice differentiable.

---

[9] With $\text{vol}_M$ we denote the canonical volume measure on $M$.

### 1.3.2. Regularity of optimal maps and the MTW condition

As we have seen in Example 1.16, even in the case of the quadratic cost we cannot expect the optimal map $T$ to be regular. However, as we have shown on Section 1.2, in case the target domain is convex any optimal map coincides with the gradient of an Aleksandrov solution to the Monge-Ampère equation and hence, by the results in Chapter 2, this is smooth if so are the densities.

A natural question is whether one may prove some partial regularity on $T$ when the convexity assumption on the support of the target domain is removed. In [52, 55] the authors proved the following result:

**Theorem 1.30.** *Let $f$ and $g$ be smooth probability densities bounded away from zero and infinity on two bounded open sets $X$ and $Y$ respectively, and let $T$ denote the unique optimal transport map from $f\mathscr{L}^n$ to $g\mathscr{L}^n$ for the quadratic cost $|x - y|^2/2$. Then there exist $X' \subset X$ and $Y' \subset Y$ open, such that $|X \setminus X'| = |Y \setminus Y'| = 0$ and $T : X' \to Y'$ is a smooth diffeomorphism.*

In the case of general cost functions on $\mathbb{R}^n$, or when $c(x, y) = d(x, y)^2/2$ on a Riemannian manifold $M$, the situation is much more complicated. Indeed, as shown by Ma, Trudinger, and Wang [80], and Loeper [78], in addition to suitable convexity assumptions on the support of the target density (or on the cut locus of the manifold when supp$(g) = M$, [61]), a very strong structural condition on the cost function, the so-called *MTW condition*, is needed to ensure even the continuity of the map. For the sake of completeness we recall it and we refer to [95, Chapter 12] for a nice presentation of its analytical and geometric consequence (in particular its relation to the *connectness* of the $c$ subdifferential, see also [78]). A $C^4$ cost $c$ is said to satisfy the MTW condition if for every $(x, y) \in X \times Y$ and every two vectors $\xi$ and $\eta$ with $\xi \cdot \eta = 0$, it holds

$$\left(c_{ij,rs} - c^{p,q}c_{ij,p}c_{q,rs}\right)c^{r,k}c^{s,l}\xi_i\xi_j\eta_i\eta_j \leq 0. \tag{1.39}$$

Here the subscripts of $c$ before the comma means derivatives in $x$, after the comma in $y$, $c^{i,j}$ is the inverse of $c_{i,j}$ and the summation is over repeated indexes.

If the MTW condition holds (together with suitable convexity assumptions on the target domain), then the optimal map is smooth [91, 92, 58, 75, 56]. On the other hand, if the MTW condition fails just at one point, then one can construct smooth densities (both supported on domains which satisfy the needed convexity assumptions) for which the optimal transport map is not continuous [78] (see also [51]).

In the case of Riemannian manifolds, the MTW condition for $c = d^2/2$ is very restrictive: indeed, as shown by Loeper [78] it implies that $M$ has non-negative sectional curvature, and actually it is much stronger than the latter [72, 60]. In particular, if $M$ has negative sectional curvature, then the MTW condition fails at every point. Let us also mention that, up to now, the MTW condition is known to be satisfied only for very special classes of Riemannian manifolds, such as spheres, their products, their quotients, and their perturbations [79, 59, 38, 73, 62, 57, 39], and for instance it is known to fail on sufficiently flat ellipsoids [60].

A natural question which arises from the above discussion is whether for a general cost function $c$ it is possible to prove a partial regularity theorem in the spirit of Theorem 1.30. In Chapter 6 we will show that this is actually the case, see Theorems 6.1 and 6.2.

# Chapter 2
# The Monge-Ampère equation

The aim of this Chapter is to introduce the main ideas behind the regularity theory of Aleksandrov solutions to the Monge-Ampère equation and to give a proof of Caffarelli $C^{1,\alpha}$ regularity theorem [18, 20]. Many of the tools developed in this Chapter will play a crucial role in the proof of the Sobolev regularity in Chapter 3. In the last Section we show, without proofs, how to build smooth solutions to the Monge-Ampère equation throughout the method of continuity.

We will investigate mainly properties of (Aleksandrov) solutions of the equation [1]

$$\begin{cases} \det D^2 u = f & \text{in } \Omega \\ u = 0 & \text{on } \partial\Omega \end{cases} \tag{2.1}$$

where $\Omega \subset \mathbb{R}^n$ is a bounded a convex set and

$$0 < \lambda \le f \le 1/\lambda \tag{2.2}$$

for some positive constant $\lambda$. More in general, since we have in mind applications to optimal transportation, we are interested in Aleksandrov solution of

$$\lambda \le \det D^2 u \le 1/\lambda \quad \text{in } \Omega. \tag{2.3}$$

As we said the aim is to prove the following theorem [2]

**Theorem 2.1 (Caffarelli [18, 20]).** *Let* $u : \Omega \to \mathbb{R}$ *be a* strictly convex *solution of (2.3). Then* $u \in C^{1,\alpha}_{\text{loc}}(\Omega)$ *for some universal* $\alpha$. *More precisely for every* $\Omega' \Subset \Omega$ *there exists a constant* $C$ *depending on* $\lambda$, $\Omega'$ *and on*

---

[1] We are identifying the function $f$ with the measure $f \mathcal{L}^n$

[2] In the sequel we are going to call *universal* any constant which depends only on $n$ and $\lambda$. Moreover we will write $a \lesssim b$ if $a/b$ is bounded from above by a universal constant, $a \gtrsim b$ if $b \lesssim a$ and $a \approx b$ if $a \lesssim b$ and $b \lesssim a$.

*the modulus of convexity of u such that*

$$\sup_{\substack{x,y\in\Omega' \\ x\neq y}} \frac{|\nabla u(x) - \nabla u(y)|}{|x - y|^\alpha} \leq C.$$

In the above Theorem we have used the notion of modulus of convexity of a function on a domain $\Omega$. Among the many (essentially equivalent) way to define it, the one which is more convenient for our scopes is the following (see (2.7) for the definition of $S(x, p, t)$): if $x_0 \in \Omega$ then the modulus of convexity of $u$ at $x_0$ is defined as

$$\omega(x_0, u, t) = \sup_{p\in\partial u(x_0)} \operatorname{diam} S(x_0, p, t). \tag{2.4}$$

A function $u$ is strictly convex at $x_0$ if $\omega(u, x_0, 0^+) = 0$ which means that every supporting plane to $u$ at $x_0$ touches the graph of $u$ only at $x_0$. The modulus of convexity of $u$ on a subdomain $\Omega'$ is then defined as

$$\omega_{\Omega'}(u, t) = \sup_{x\in\Omega'} \omega(x, u, t). \tag{2.5}$$

It is easy to see that if $\Omega' \Subset \Omega$ and $u$ is strictly convex at every point $x \in \Omega'$ then $\omega_{\Omega'}(u, 0^+) = 0$.

In order to apply the above result (and the ones of Chapter 3) to optimal transport we need to prove the strict convexity of solutions.

**Theorem 2.2 (Caffarelli [21]).** *Let us assume that $\mu = \rho_1 \mathscr{L}^n$ and $\nu = \rho_2 \mathscr{L}^n$ satisfy the assumption at the beginning of Section 1.2 and that $\Omega_2$ is convex. If $T = \nabla u$ is the optimal transport between $\mu$ and $\nu$, then $u$ is strictly convex inside $\Omega_1$. Moreover the modulus of strict convexity depends only on $\lambda$, $\Omega_1$ and $\Omega_2$ and is bounded as soon as $\Omega_2$ varies in a compact class with respect to the Hausdorff distance.*

## 2.1. Aleksandrov maximum principle

We start recalling the Aleksandrov maximum principle which is a key tool in the study of the Monge-Ampère equation and, more in general, of fully nonlinear elliptic PDE (see [25, 66]).

**Lemma 2.3.** *Let $u$ and $v$ be convex functions in $\mathbb{R}^n$. If $A$ is an open and bounded set such that $u = v$ on $\partial A$ and $u \leq v$ in $A$, then*

$$\partial u(A) \supset \partial v(A). \tag{2.6}$$

*In particular $\mu_u(A) \geq \mu_v(A)$.*

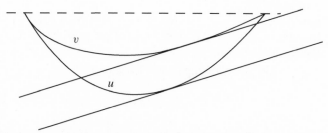

**Figure 2.1.** Moving down a supporting plane to $v$ and lifting it up until it touches $u$ we obtain a supporting place to $u$ at some point $\bar{x} \in A$.

*Proof.* Let $p \in \partial v(x)$ for some $x \in U$, this means that the plane

$$y \mapsto v(x) + p \cdot (y - x)$$

is a supporting plane to $v$ at $x$. Moving this plane down and lifting it up until it touches the graph of $u$ for the first time we see that, for some constant $a \leq v(x)$,

$$y \mapsto a + p \cdot (y - x)$$

is a supporting plane to $u$ at some point $\bar{x} \in \overline{A}$, see Figure 2.1.

Since $u = v$ on $\partial A$ we see that, if $\bar{x} \in \partial A$, $a = v(x)$ and thus $u(x) = v(x)$ and the plane is also supporting $u$ at $x$. In conclusion $p \in \partial u(A)$, proving the inclusion (2.6). $\qquad\square$

**Theorem 2.4 (Aleksandrov maximum principle).** *Let $u : \Omega \to \mathbb{R}$ be a convex function defined on an open, bounded and convex domain $\Omega$. If $u = 0$ on $\partial \Omega$, then*

$$|u(x)|^n \leq C_n (\operatorname{diam} \Omega)^{n-1} \operatorname{dist}(x, \partial \Omega) |\partial u(\Omega)| \qquad \forall x \in \Omega,$$

*here $C_n$ is a geometric constant depending only on the dimension.*

*Proof.* Let $(x, u(x))$ be a point on the graph of $u$ and let us consider the cone $C_x(y)$ with vertex on $(x, u(x))$ and base $\Omega$, that is the graph of one-homogeneous function (with respect to dilatation with center $x$) which is 0 on $\partial \Omega$ and equal to $u(x)$ at $x$. Since by convexity $u(y) \geq C_x(y)$, Lemma 2.3 implies

$$|\partial C_x(x)| \leq |\partial C_x(\Omega)| \leq |\partial u(\Omega)|.$$

To conclude the proof we have only to show that

$$|\partial C_x(x)| \geq \frac{|u(x)|^n}{C_n (\operatorname{diam} \Omega)^{n-1} \operatorname{dist}(x, \partial \Omega)}.$$

Let $p$ such that $|p| < |u(x)|/\operatorname{diam}\Omega$ and let us consider a plane with slope $p$, moving it down and lifting it up until it touches the graph of $C_x$ we see that it has to be supporting to some point $\bar{y} \in \Omega$. Since $C_x$ is a cone it also has to be supporting at $x$. This means

$$\partial C_x(x) \supset B(0, |u(x)|/\operatorname{diam}\Omega).$$

Let now $\bar{x} \in \partial\Omega$ be such that $\operatorname{dist}(x, \partial\Omega) = |x - \bar{x}|$ and let $q$ be a vector with the some direction of $(\bar{x} - x)$ and with modulus less then $|u(x)|/\operatorname{dist}(x, \partial\Omega)$, then the plane $u(x) + q \cdot (y - x)$ will be supporting $C_x$ at $x$ (see Figure 2.2), that is

$$q := \frac{\bar{x} - x}{|\bar{x} - x|} \frac{|u(x)|}{|\operatorname{dist}(x, \partial\Omega)} \in \partial C_x(x).$$

By the convexity of $\partial C_x(x)$ we have that it contains the cone $\mathcal{C}$ generated by $q$ and $B(0, |u(x)|/\operatorname{diam}\Omega)$. Since, for some geometric constant $C_n$,

$$|\mathcal{C}| = \frac{|u(x)|^n}{C_n(\operatorname{diam}\Omega)^{n-1}\operatorname{dist}(x, \partial\Omega)},$$

we conclude the proof. $\qquad\square$

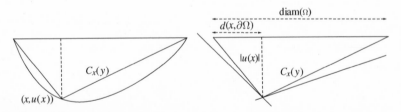

**Figure 2.2.** Every plane with slope $|p| \leq |u(x)|/\operatorname{diam}(\Omega)$ supports the graph of $C_x$ at $x$. Moreover there exists a supporting plane with slope $|q| \leq |u(x)|/\operatorname{dist}(x, \partial\Omega)$.

Another consequence of Lemma 2.3 is the following comparison principle

**Lemma 2.5.**   *Let $u, v$ be convex functions defined on a open and bounded convex set $\Omega$. If $u \geq v$ on $\partial\Omega$ and (in the sense of Monge-Ampère measures)*

$$\det D^2 u \leq \det D^2 v \quad in \ \Omega,$$

*then $u \geq v$ in $\Omega$.*

*Proof.* Up to substituting $u$ with $u + \varepsilon$ and send $\varepsilon \to 0$ at the end of the proof we can assume that

$$\inf_{\partial\Omega}(u - v) \geq \varepsilon.$$

Let us assume that $u(\bar{x}) < v(\bar{x})$ for some $\bar{x} \in \Omega$ and define $v_\delta = v + \delta|x - \bar{x}|^2$. Choosing $\delta \ll \varepsilon$ we see that

$$A := \{u < v_\delta\} \Subset \Omega,$$

but Lemma 2.3 yields

$$|\partial u(A)| \geq |\partial v_\delta(A)| \geq |\partial v(A)| + 2\delta|A| > |\partial u(A)|,$$

a contradiction. Here we have used that for two convex functions $w$ and $v$

$$\det(D^2 w + D^2 v) \geq \det D^2 w + \det D^2 v$$

as measures. To prove it, notice that thanks to Proposition 1.25 and an approximation argument we can reduce to the case in which $w$ and $v$ are smooth, but then it follows from the inequality $\det(M + N) \geq \det M + \det N$, which holds for any two positive matrices $M$ and $N$.   $\square$

The above Lemma implies, in particular, that solutions to (2.1) are unique.

## 2.2. Sections of solutions of the Monge-Ampère equation and Caffarelli regularity theorems

One of the main features of the Monge-Ampère equation is its affine invariance: if we right compose a solution of (2.3) with an affine transformation of determinant 1 it is immediate to see that the new function is still a solution of (2.3). Due to this invariance it is impossible to have estimates which do not depend on the geometry of the domain (see section 2.3). In spite of this difficulty, we will see how the affine invariance of the equation can be used in order to deduce properties of the solution.

A key role in the study of solutions of (2.3) is played by the *sections* of $u$, which play for the Monge-Ampère equation the same role that balls play for an uniformly elliptic equation.

Given $u : \Omega \to \mathbb{R}$ a convex function, for any point $x$ in $\Omega$, $p \in \partial u(x)$, and $t \geq 0$, we define the section centered at $x$ with height $t$ (with respect to $p$) as

$$S(x, p, t) := \{y \in \Omega : u(y) \leq u(x) + p \cdot (y - x) + t\}, \qquad (2.7)$$

see Figure 2.3.

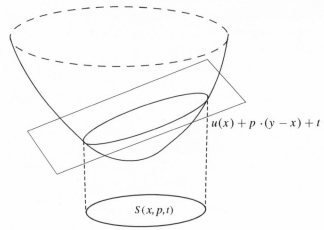

**Figure 2.3.** Section of a convex function $u$.

We say that an open bounded convex set $Z \subset \mathbb{R}^n$ is *normalized* if

$$B(0, 1) \subset Z \subset B(0, n).$$

By John's Lemma (see Appendix B), for every open bounded convex set $S$ there exists an ellipsoid $E$ such that

$$E \subset S \subset nE,$$

where the dilation is respect to the center of $E$. Hence there exists an (invertible) orientation preserving affine transformation $T : \mathbb{R}^n \to \mathbb{R}^n$ such that $T(S)$ is normalized. In particular

$$\frac{|B_1|}{|S|} \leq \det T \leq \frac{n^n|B_1|}{|S|}. \tag{2.8}$$

Notice that in the sequel we are not going to notationally distinguish between an affine transformation and its linear part, since it will always be clear to what we are referring to. In particular, we will use the notation

$$\|T\| := \sup_{|v|=1} |Av|, \qquad Tx = Ax + b. \tag{2.9}$$

One useful property which we will use is the following identity: if we denote by $T^*$ the adjoint of $T$, then

$$\|T^*T\| = \|T^*\|\|T\|. \tag{2.10}$$

(This can be easily proved using the polar decomposition of matrices.)

Whenever $u$ is a strictly convex solution of (2.3) for any $x \in \Omega' \Subset \Omega$ one can choose $t_0 > 0$ sufficiently small (depending only on $\Omega'$ and the modulus of convexity of $u$) so that $S(x, p, t) \Subset \Omega$ for all $t < t_0$. Then, if $T$ is the affine transformation which normalizes $S(x, p, t)$, the function

$$v(z) := (\det T)^{2/n} \left[ u(T^{-1}z) - u(x) - p \cdot (T^{-1}z - x) - t \right], \quad p \in \partial u(x), \tag{2.11}$$

solves

$$\begin{cases} \lambda \leq \det D^2 v \leq 1/\lambda & \text{in } Z, \\ v = 0 & \text{on } \partial Z, \end{cases} \tag{2.12}$$

with $Z := T(S(x, p, t))$ renormalized. We are going to call $v$ a *normalized solution*.

**Lemma 2.6.** *Let $v$ be a normalized solution, then there exist universal positive constants $c_1$ and $c_2$ such that*

$$0 < c_1 \leq \left| \inf_Z v \right| \leq c_2. \tag{2.13}$$

*Proof.* Consider the functions $w_1 = \lambda(|x|^2 - 1)/2$ and $w_2 = (|x|^2 - n^2)/2\lambda$ and apply Lemma 2.5 to see that $w_2 \leq v \leq w_1$.  $\square$

**Remark 2.7.** Since $|\inf_Z v| \approx 1$ we see that

$$t = \inf_{y \in S(x,p,t)} |u(y) - u(x) - p \cdot (y - x)| = \frac{|\inf_Z v|}{|\det T|^{2/n}} \approx \frac{1}{|\det T|^{2/n}}$$

which, together with (2.8) implies $|S(x, p, t)| \approx t^{n/2}$.

We now show that in case a solution of (2.3) is not strictly convex, then the set where it coincides with one of its supporting plane has to cross the domain of definition.

Recall that if $C$ is a closed convex set a point $\bar{x} \in C$ is said extremal if it cannot be expressed as non trivial convex combination of two points in $C$ or, equivalently, if $C \setminus \{\bar{x}\}$ is convex. We recall the following Lemma about extreme points.

**Lemma 2.8.** *A point $\bar{x}$ is extremal for $C$ if and only if for every $\delta > 0$ there exists a closed half space $H$ such that $\bar{x} \in \text{int} H$ and $C \cap H \subset B(\bar{x}, \delta)$.*

*Proof.* We are going to prove only the implication that we need. Let $B = \text{co}[C \setminus B(\bar{x}, \delta)]$. Since $\bar{x}$ is extremal for $\delta$ small $B$ is a non empty convex set such that $\bar{x} \notin B$. Then there exists a closed hyperplane $K$ strongly separating $B$ and $\bar{x}$. If $H$ is the closed half space bounded by $K$ which contains $\bar{x}$ in its interior, then $H \cap C \subset B(\bar{x}, \delta)$.  $\square$

We are now ready to prove the following theorem, due to Caffarelli. The strategy of the proof is the following: in case the set where $u$ coincides with one of its supporting planes $\ell$ has an extreme point in $\Omega$, we cut the graph of $u$ with suitable linear functions $\tilde{\ell}_\varepsilon$ converging to $\ell$. If we look at the sets $K_\varepsilon = \{u - \ell_\varepsilon \leq 0\}$ and to their renormalization $K_\varepsilon^*$, we see that, from one side, by Aleksandrov maximum principle, the minimum of $v_\varepsilon$ (the renormalization of $u_\varepsilon$) should stay far from the boundary, and from the other side it converges to $\partial K_\varepsilon^*$.

**Theorem 2.9 (Caffarelli [18]).** *Let $u$ be a solution of (2.3) inside a convex set $\Omega$ and let $\ell(x)$ a supporting slope to $u$ at some point $x \in \Omega$. If the convex set*

$$W = \{x \in \Omega : \quad u(x) = \ell(x)\}$$

*contains more than one point, then it cannot have extremal points in $\Omega$.*

*Proof.* Up to subtracting a linear function we can assume that $\ell = 0$, $u \geq 0$ and, by contradiction, that the set $W = \{u = 0\}$ has a extremal point in $\Omega$. Using Lemma 2.8 we can thus assume that (up to a rotation of coordinates) $K_0 := W \cap \{x_1 \geq 0\}$ is non empty and compactly supported in $\Omega$. Moreover, always thanks to the above mentioned lemma,

$$x^0 := \operatorname*{argmax}_{K_0} x_1$$

satisfies $x_1^0 > 0$. Let us define the convex domains

$$K_\varepsilon = \{u \leq \varepsilon x_1\} \cap \{x_1 \geq 0\}.$$

Notice that $\cap_\varepsilon K_\varepsilon = K_0$ and that

$$x^\varepsilon := \operatorname*{argmax}_{K_\varepsilon} x_1 \to x^0$$

as $\varepsilon \to 0$. Let $w_\varepsilon(x) = u(x) - \varepsilon x_1$, then

$$\begin{cases} \lambda \leq \det D^2 w_\varepsilon \leq 1/\lambda & \text{in } K_\varepsilon \\ w_\varepsilon = 0 & \text{on } \partial K_\varepsilon. \end{cases}$$

Let us construct the normalized solutions $v_\varepsilon$ on the normalized sets $K_\varepsilon^* = T_\varepsilon(K_\varepsilon)$ as in (2.12). Let us look to the points $T_\varepsilon(x^0)$. First notice that (recall (2.11) and that $u$ is positive)

$$\frac{v_\varepsilon(T_\varepsilon(x^0))}{\inf_{K_\varepsilon^*} v_\varepsilon} = \frac{w_\varepsilon(x^0)}{\inf_{K_\varepsilon} w_\varepsilon} \geq \frac{\varepsilon x_1^0}{\varepsilon \max_{K_\varepsilon} x^1} = \frac{x_1^0}{x_1^\varepsilon} \to 1,$$

since $x^\varepsilon \to x^0$. Hence, for $\varepsilon$ small, $v_\varepsilon(T_\varepsilon(x^0)) \approx \inf_{K_\varepsilon^*} v_\varepsilon \approx 1$, by Lemma 2.6. Thanks to Theorem 2.4 and to the fact that $K_\varepsilon^*$ is normalized we deduce

$$\text{dist}(T_\varepsilon(x_0), \partial K_\varepsilon^*) \gtrsim 1. \tag{2.14}$$

If we now consider the three parallel hyperplanes $\Pi_1 = \{x_1 = 0\}$, $\Pi_2 = \{x_1 = x_1^0\}$ and $\Pi_3 = \{x_1 = x_1^\varepsilon\}$, we see that

$$\frac{\text{dist}(\Pi_2, \Pi_3)}{\text{dist}(\Pi_1, \Pi_3)} = \frac{x_1^\varepsilon - x_1^0}{x_1^\varepsilon} \to 0.$$

Since the above ratio is affine invariant we get

$$\frac{\text{dist}(T_\varepsilon(\Pi_2), T_\varepsilon(\Pi_3))}{\text{dist}(T_\varepsilon(\Pi_1), T_\varepsilon(\Pi_3))} \to 0.$$

Now, $T_\varepsilon(\Pi_1)$ and $T_\varepsilon(\Pi_3)$ are supporting planes to $K_\varepsilon^*$ and thus their distance is bounded from above by the diameter of $K_\varepsilon^*$ which, since $K_\varepsilon^*$ is normalized, is bounded by $2n$. In conclusion

$$\text{dist}(T_\varepsilon(x_0), \partial K_\varepsilon^*) \leq \text{dist}(T_\varepsilon(\Pi_2), T_\varepsilon(\Pi_3)) \to 0,$$

contradicting (2.14). □

The above theorem says that one of the following two alternatives holds: either $u$ is strictly convex or the set where $u$ coincides with one of its tangent planes has to cross its domain of definition. This latter fact can actually happen if $n \geq 3$ (while if $n = 2$ any solution of $\det D^2 u \geq \lambda$ is strictly convex, see [23]) as it is shown by the function

$$w(x) = (1 + (x_1)^2)|x'|^{2-2/n} \qquad x = (x_1, x'),$$

which is a convex generalized solution of (2.3) in a sufficiently small ball around the origin (see [67, Section 5.5]).

We are now in the position to prove Theorem 2.1; by localization (which is possible due to the strict convexity of $u$) it is enough to show that a normalized solution is $C^{1,\alpha}_{\text{loc}}$. A first step is the following Lemma.

**Lemma 2.10 (The class of normalized solution is compact).** *Let $Z_k$ be a sequence of normalized domains and $v_k$ be a sequence of normalized solutions of (2.12) defined on $Z_k$. Then, up to subsequence, there exists a limiting function $v_\infty$ and a limiting normalized domain $Z_\infty$ such that $Z_k \to Z_\infty$ in the Hausdorff metric and $v_k$ to $v_\infty$ locally uniformly in $Z_\infty$. In particular $v_\infty$ is a normalized solution.*

*Proof.* By classical theorems, up to subsequences, the sets $Z_k$ converge to a limiting normalized convex set, $Z_\infty$. Since $\operatorname{osc} v_k \approx 1$ (by Lemma 2.6), convexity implies that if $K \subset Z_k$ then

$$\operatorname{Lip}(v_k, K) \lesssim \frac{1}{\operatorname{dist}(Z_k, K)}.$$

Since every compact set contained in $Z_\infty$ is contained in $Z_k$ for $k$ large enough, we see that a subsequence of $\{v_k\}_{k\in\mathbb{N}}$ locally uniformly converges to a convex function $v_\infty$ which is a solution of (2.3) thanks to Proposition 1.25. To fix the boundary data notice that, by Theorem 2.4,

$$-\operatorname{dist}^{1/n}(\cdot, \partial Z_k) \lesssim v_k \leq 0 \quad \text{in } Z_k$$

and pass to limit as $k$ goes to infinity. □

**Lemma 2.11.** *For any normalized solution $v$ on a normalized domain $Z$, the modulus of strict convexity of $v$ on $Z' \Subset Z$ depends only on $\operatorname{dist}(Z', \partial Z)$, $\lambda$ and $n$. More precisely, there exists a function $\omega$ depending only on the previous mentioned quantities such that $\omega(0^+) = 0$ and*

$$\sup_{x\in Z'} \sup_{p\in\partial u(x)} \operatorname{diam}(S(x, p, t)) \leq \omega(t).$$

*Proof.* Assume that this is not the case, then for $\varepsilon_0 > 0$ there exists a sequence of normalized domains $Z_k$, of normalized solutions $\{v_k\}_{k\in\mathbb{N}}$, of points $x_k \in Z_k'$, $y_k \in Z_k$ such that $|x_k - y_k| \geq \varepsilon_0$, $\operatorname{dist}(x_k, \partial Z_k) \geq \delta$,

$$u_k(y_k) \leq u_k(x_k) + p_k \cdot (y_k - x_k) + \frac{1}{k} \qquad p_k \in \partial u_k(x_k).$$

Since $\operatorname{dist}(x_k, \partial Z_k) \geq \delta$ and $\operatorname{osc} v_k \approx 1$, by convexity,

$$|p_k| \lesssim \frac{1}{\delta}.$$

With the aid of Lemma 2.10 we find a limiting normalized solution $v_\infty$ defined on a normalized domain $Z_\infty$ and limiting points $x_\infty$, $y_\infty$ in $Z_\infty$, $p_\infty \in \partial v_\infty(x_\infty)$ such that $|x_\infty - y_\infty| \geq \varepsilon_0$, $\operatorname{dist}(x_\infty, \partial Z_\infty) \geq \delta$ and

$$u_\infty(y_\infty) \leq u_\infty(x_\infty) + p_\infty \cdot (y_\infty - x_\infty).$$

But then the set where $v_\infty$ coincides with its supporting plane

$$\ell(y) = u_\infty(x_\infty) + p_\infty \cdot (y - x_\infty)$$

has more than one point and thus, by Theorem (2.9), it has to cross $Z_\infty$. This contradicts the fact that $v_\infty = 0$ on $\partial Z_\infty$ and $\inf v_\infty \approx -1$. □

We are now ready to prove Theorem 2.1, as we said it will be enough to show that normalized solution are $C^{1,\alpha}$. The strategy of the proof goes as follows. By Theorem 2.9 we know that $v$ is strictly convex, in particular, if we consider the cone $C_\beta \subset \mathbb{R}^{n+1}$ with vertex the minimum point of $v$, $(x_0, v(x_0))$, and base $\{v = (1 - \beta) \min v\} \times \{(1 - \beta) \min v\}$ then $C_{1/2}$ is strictly below $C_1$, see Figure 2.4. By a contradiction compactness argument we see that there exists a universal $\delta_0 < 1$ such that, if $C_\beta$ is the graph of $h_\beta$,

$$h_{1/2} \le (1 - \delta_0)h_1.$$

Rescaling and iterating we show that $v$ is $C^{1,\alpha}$ at the minimum. Since, up to subtracting a linear function, every point behaves like a minimum point, $v$ is $C^{1,\alpha}$ at every point.

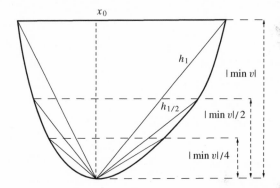

**Figure 2.4.** The function $v$ looks flatter and flatter near its minimum $x_0$.

*Proof of Theorem* 2.1.  We divide the proof in several steps.

*Step* 1. *Section corresponding to different heights are "far away"* Let us consider

$$S_{1/2} = \left\{ v \le \min_Z v/2 \right\},$$

then $\operatorname{dist}(S_{1/2}, \partial Z) \ge 1/C$ for some universal constant $C$. This follows by Aleksandrov maximum principle, Theorem 2.4. Indeed if $x \in S_{1/2}$

$$\operatorname{dist}^{1/n}(x, \partial Z) \gtrsim -v(x) \ge -\min_Z v/2 \gtrsim 1$$

by Lemma 2.6.

*Step* 2.  Let $x_0$ be the minimum point of $v$ in $Z$ and let us consider the function $h_\beta$ whose graph is the cone generated by $\{v = (1 - \beta) \min v\} \times$

$\{(1 - \beta) \min v\}$ and the point $(x_0, v(x_0))$. Then there exists a universal constant $\delta_0$ such that

$$h_{1/2} \le (1 - \delta_0)h_1.$$

Notice that since the functions $h_\beta$ are 1-homogeneous with respect to dilation with center $x_0$ it is enough to prove the above inequality for all points $x \in \partial S_{1/2}$.

As we said proof is by contraddiction-compactness. Suppose that the claim is false, then we find normalized solutions $v_k$ defined on normalized domains $Z_k$ and points $x_k \in \partial S_{1/2}^k$ such that

$$h_1^k(x_k) \ge \left(1 - \frac{1}{k}\right)h_{1/2}^k(x_k).$$

Taking into account that

$$\mathrm{dist}(S_{1/2}^k, \partial Z_k) \ge 1/C$$

we deduce that the Lipschitz constants of $h_{1/2}^k$, $h_1^k$ are universally bounded (recall the $h_\beta^k$ are 1-homogeneous functions). Thus with the aid of Lemma 2.10 we find a limiting normalized solution $v_\infty$ defined on a normalized domain $Z_\infty$ and a point $x_\infty \in \partial S_{1/2}^\infty$, such that

$$h_1^\infty(x_\infty) = h_{1/2}^\infty(x_\infty).$$

But then (by homogeneity) the above equality is true on the segment through $x_0$ (the minimum point of $v_\infty$) and $x_\infty$. This means that the graph of $v_\infty$ contains a segment and thus coincides with one of its supporting planes in more than one point, but then this set has to cross $Z_\infty$ and we get a contradiction as in the proof of Lemma 2.11.

*Step 3. $v$ is $C^{1,\alpha}$ at the minimum point.* By Step 2 we know that

$$h_{1/2} \le (1 - \delta_0)h_1.$$

We now consider the set $S_{1/2}$, its normalization $\tilde{Z} = T(S_{1/2})$ and the function $\tilde{v} = (\det T)^{2/n} v \circ T^{-1}$. Then $\tilde{v}$ is a normalized solution and thus

$$\tilde{h}_{1/2} \le (1 - \delta_0)\tilde{h}_1,$$

where $\tilde{h}_\beta$ are defined as in Step 2 starting from $\tilde{v}$. Coming back to $v$

$$h_{1/4} \le (1 - \delta_0)^2 h_1,$$

and thus, iterating,

$$h_{2^{-k}} \le (1 - \delta_0)^k h_1,$$

see Figure 2.4. Since the domain $Z = \{h_1 \leq 0\}$ is normalized and $h_1$ is 1-homogeneous it easily follows that

$$h_1(x) \leq C_0|x - x_0| + v(x_0),$$

where $x_0$ is the minimum point of $v$ and $C_0$ is universal. Since $v \leq h_{2^{-k}}$ as long as $h_{2^{-k}} \leq (1 - 2^{-k}) v(x_0)$, we see that if we define $\alpha'$ through the relation $(1 - \delta_0) = 2^{-\alpha'}$ and for every $x$ we define $k$ such that

$$(-C_0/v(x_0))2^{-(k+1)(1-\alpha')} \leq |x - x_0| \leq (-C_0/v(x_0))2^{-k(1-\alpha')},$$

then

$$h_{2^{-k}}(x) \leq (1 - 2^{-k}) v(x_0).$$

Hence, for $\alpha = \alpha'/(1 - \alpha')$,

$$v(x) - v(x_0) \leq (-v(x_0))2^{-k}$$

$$= 2(-v(x_0)) \left(\frac{-v(x_0)}{C_0}\right)^{1/(1-\alpha')} \left(\frac{C_0 2^{-(k+1)(1-\alpha')}}{-v(x_0)}\right)^{1/(1-\alpha')}$$

$$\leq C_1|x - x_0|^{1+\alpha},$$

where

$$C_1 = 2(-v(x_0)) \left(\frac{C_0}{-v(x_0)}\right)^{1/(1-\alpha')}$$

is universal since $-v(x_0) = -\min v \approx 1$.

*Step* 4, $u$ is $C^{1,\alpha}_{\text{loc}}$. It is enough to show that if $x_0 \in Z$ is such that $\text{dist}(x_0, \partial Z) \geq \delta$ then there exists a constant $C_\delta$ such that for all supporting planes $\ell_{x_0}$ at $x_0$

$$\sup_{B(x_0,r)} |v - \ell_{x_0}| \leq C_\delta r^{1+\alpha} \qquad \forall r \leq \delta/C_\delta.$$

Indeed it is well known that this will imply that $v$ is $C^{1,\alpha}$ on $Z_\delta := \{x \in Z : \text{dist}(x_0, \partial Z) \geq \delta\}$, see for instance [42, Lemma 3.1].

Let $x_0$ as above, by Lemma 2.11 we know that there exists an $\varepsilon_0 = \varepsilon_0(\delta)$ such that $\text{diam}(S(x_0, p, \varepsilon_0)) \leq \delta/2$. We normalize $S(x_0, p, \varepsilon_0) \subset Z_\delta$ and construct the normalized solution

$$w(x) = (\det T)^{2/n} \big(v(T^{-1}x) - p \cdot (T^{-1}x - x_0) - \varepsilon_0\big)$$

By the previous Step

$$|w(Tx) - w(Tx_0)| \leq C_1|Tx - Tx_0|^{1+\alpha}.$$

Coming back to $v$ we deduce

$$|v(x)-\ell_{x_0}(x)| \leq \frac{C_1}{(\det T)^{2/n}}|Tx-Tx_0|^{1+\alpha}, \quad \ell_{x_0}(x) = v(x_0)+p\cdot(x-x_0).$$

Thanks to Remark 2.7 we know that $\det T \approx 1/\varepsilon_0(\delta)^{n/2}$, and thus to prove the claim we only have to show that

$$\|T\| \leq C_\delta.$$

Since $\mathrm{diam}(S(x_0, p, \varepsilon_0)) \leq \delta/2$ and $T((S(x_0, p, \varepsilon_0))$ is normalized we see that image of a ball of radius $\delta$ through $T$ contains a ball of radius 1. Since we can assume without loss of generality that $T$ is symmetric[3] this means that the smallest eigenvalue of (the linear part of) $T$ is bounded from below by $1/\delta$. Since $\det T$ is bounded by a constant depending on $\delta$ we see that also the largest eigenvalue of $T$ is bounded by a constant depending on $\delta$. □

In order to apply the above theorem to optimal transportation, we have to show that solutions arising from such problem are strictly convex. The strategy of the proof is the same of the one of Theorem 2.9.

*Proof of Theorem* 2.2. By Proposition 1.23 we know that $u$ solves

$$\lambda^2 \mathbf{1}_{\Omega_1} \leq \det D^2 u \leq \frac{1}{\lambda^2}\mathbf{1}_{\Omega_1} \qquad \text{in } \mathbb{R}^n \tag{2.15}$$

in the Aleksandrov sense and we want to show that $u$ is strictly convex in $\Omega_1$. Let us assume it is not, then up to subtracting a linear function we can assume that $u \geq 0$ and that $W = \{u = 0\}$ has more than one point in $\Omega_1$. Let us consider the following cases:

(i) $W$ has no exposed[4] point, therefore it contains a line.
(ii) There exists an exposed points $\bar{x}$ of $W$.

In case (i) we can assume without loss of generality that $u(x_1, 0) = 0$ for all $x_1 \in \mathbb{R}$. Then if $x \in \mathbb{R}^n$ and $p \in \partial u(x)$

$$0 = u(tx_1, 0) \geq u(x_1, x') + (t-1)p_1x_1 + p'\cdot x' \geq (t-1)p_1x_1 + p'\cdot x'.$$

---

[3] Recall the polar factorization theorem for matrices which asserts that any matrix can be decomposed as $T = SO$ where $S$ is symmetric and $O$ is orthogonal. Incidentally we notice that this can be proved with the aid of Brenier Polar Factorization, Theorem 1.11.

[4] A point $\bar{x} \in \overline{W}$ is said *exposed* if there exists a closed half space $H$ such that $W \subset H$ and $\overline{W} \cap \partial H = \{\bar{x}\}$.

Sending $t \to \pm\infty$ we see that $p_1 = 0$ and thus $\partial u(\mathbb{R}^n) \subset \{p_1 = 0\}$, contradicting (2.15).

Let us consider case (ii). By Theorem 2.9 we see that $\bar{x}$ cannot belong to $\Omega_1$. Up to a change of coordinates we can assume that $\bar{x} = 0$ and

$$W \subset \{x_1 \leq 0\}, \qquad \{x_1 = 0\} \cap W = \{0\}.$$

Let us now consider two further cases:

(a) $0 \in \mathbb{R}^n \setminus \overline{\Omega}_1$,
(b) $0 \in \partial\Omega_1$.

The first case cannot happen, indeed if we consider, for $\varepsilon$ small, the functions

$$w_\varepsilon = u - \varepsilon(x_1 + \varepsilon)$$

and the sets $K_\varepsilon = \{w_\varepsilon \leq 0\} \Subset \mathbb{R}^n \setminus \overline{\Omega}_1$, we see that (2.15) implies $|\partial w_\varepsilon(K_\varepsilon)| = 0$, while

$$\inf_{K_\varepsilon} w_\varepsilon < 0,$$

contradicting Aleksandrov maximum principle, Theorem 2.4. Let us assume now that we are in case (b), that is

$$0 \in \partial\Omega_1, \qquad \Omega_1 \cap W \ni \tilde{x} \neq 0, \qquad W \subset \{x_1 \leq 0\}.$$

Let us consider the functions

$$v_\varepsilon = u - \varepsilon(x_1 - (\tilde{x}_1 - \varepsilon))$$

and the sets $J_\varepsilon = \{v_\varepsilon \leq 0\}$. Notice that for $\varepsilon$ small, $J_\varepsilon$ are open and bounded convex sets and $\tilde{x} \in J_\varepsilon$, in particular $|J_\varepsilon \cap \Omega_1| > 0$. Let $T_\varepsilon$ be the transformation which normalizes $J_\varepsilon$ and let $\tilde{v}_\varepsilon = (\det T_\varepsilon)^{2/n} v_\varepsilon \circ T_\varepsilon^{-1}$, then

$$\lambda^2 \mathbf{1}_{T_\varepsilon(\Omega_1)} \leq \det D^2 \tilde{v}_\varepsilon \leq \frac{1}{\lambda^2} \mathbf{1}_{T_\varepsilon(\Omega_1)} \qquad \text{in } \mathbb{R}^n.$$

Since $T_\varepsilon(J_\varepsilon)$ is normalized Aleksandrov maximum principle, Theorem 2.4, implies

$$|\tilde{v}_\varepsilon(T_\varepsilon(0))|^n \leq C(n)|T_\varepsilon(J_\varepsilon) \cap T_\varepsilon(\Omega_1)| \operatorname{dist}(T_\varepsilon(0), T_\varepsilon(\partial J_\varepsilon))/\lambda^2.$$

Since $\det T_\varepsilon \approx 1/|J_\varepsilon|$, the above estimates translate in

$$|v_\varepsilon(0)|^n \leq C(n, \lambda)|J_\varepsilon||J_\varepsilon \cap \Omega_1| \operatorname{dist}(T_\varepsilon(0), T_\varepsilon(\partial J_\varepsilon)). \qquad (2.16)$$

On the other side, if we consider the dilation with respect to 0 of $T_\varepsilon(J_\varepsilon)$, we see that for all $x \in T_\varepsilon(J_\varepsilon)/2$, $p \in \partial u(x)$,

$$|p| \leq \frac{|\inf_{T_\varepsilon(J_\varepsilon)} \tilde{v}_\varepsilon|}{\operatorname{dist}\big(T_\varepsilon(J_\varepsilon)/2, T_\varepsilon(\partial J_\varepsilon)\big)} \leq C(n) \left| \inf_{T_\varepsilon(J_\varepsilon)} \tilde{v}_\varepsilon \right|.$$

where we have used that $\operatorname{dist}\big(T_\varepsilon(J_\varepsilon)/2, T_\varepsilon(\partial J_\varepsilon)\big) \geq 1/C(n)$ since $T_\varepsilon(J_\varepsilon)$ is normalized. Thus

$$\partial \tilde{v}_\varepsilon(T_\varepsilon(J_\varepsilon)/2) \subset B\left(0, \left|C(n) \inf_{T_\varepsilon(J_\varepsilon)} \tilde{v}_\varepsilon\right|\right). \tag{2.17}$$

Hence

$$\frac{\lambda^2}{2^n}|T_\varepsilon(J_\varepsilon) \cap T_\varepsilon(\Omega_1)| = \lambda^2 |T_\varepsilon(J_\varepsilon)/2 \cap T_\varepsilon(\Omega_1)/2|$$

$$\leq \lambda^2 |T_\varepsilon(J_\varepsilon)/2 \cap T_\varepsilon(\Omega_1)|$$

$$\leq |\partial \tilde{v}_\varepsilon(T_\varepsilon(J_\varepsilon)/2)| \leq C(n) \left| \inf_{T_\varepsilon(J_\varepsilon)} \tilde{v}_\varepsilon \right|^n,$$

where in the last step we have used (2.17). Back to $v_\varepsilon$:

$$|J_\varepsilon||J_\varepsilon \cap \Omega_1| \leq C(n, \lambda) \left| \inf_{J_\varepsilon} v_\varepsilon \right|^n. \tag{2.18}$$

Since $|J_\varepsilon||J_\varepsilon \cap \Omega_1| > 0$, we see that

$$\frac{|v_\varepsilon(0)|^n}{|\inf_{J_\varepsilon} v_\varepsilon|^n} \leq C(n, \lambda) \operatorname{dist}(T_\varepsilon(0), T_\varepsilon(\partial J_\varepsilon)).$$

Now arguing as in the proof of Theorem 2.9 we get a contradiction since

$$\frac{|v_\varepsilon(0)|}{|\inf_{J_\varepsilon} v_\varepsilon|} \to 1 \quad \text{and} \quad \operatorname{dist}(T_\varepsilon(0), T_\varepsilon(\partial J_\varepsilon)) \to 0$$

as $\varepsilon \to 0$.

The fact that the modulus of strict convexity is uniformly bounded as $\Omega_2$ varies in a compact class of convex sets follows by a simple contradiction compactness argument (cp. the proof of Lemma 2.11). $\qquad\Box$

We conclude this section proving some properties of sections of solutions of the Monge-Ampère equation on a domain $\Omega$. These properties show that $\Omega$ endowed with the Lebesgue measure and the family of "balls" $\{S(x, p, t)\}_{x \in \Omega, t \in \mathbb{R}}$ is a space homogenous type in the sense of Coifman and Weiss, see [36, 68]. This will play a key role in the proof of Sobolev regularity in Chapter 3.

Since, by Theorem 2.1, $u$ is continuously differentiable $\partial u(x)$ reduces to $\{\nabla u(x)\}$, and in the sequel we will simply write $S(x, t)$ for $S(x, \nabla u(x), t)$. Moreover, given $\tau > 0$, we will use the notation $\tau S(x, t)$ to denote the dilation of $S(x, t)$ by a factor $\tau$ with respect to $x$, that is

$$\tau S(x, t) := \left\{ y \in \mathbb{R}^n : x + \frac{y - x}{\tau} \in S(x, t) \right\}. \qquad (2.19)$$

**Proposition 2.12.** *Let $u$ be a convex Aleksandrov solution of* (2.1) *with $0 < \lambda \leq f \leq 1/\lambda$. Then, for any $\Omega' \Subset \Omega'' \subset \Omega$, there exists a positive constant $\rho = \rho(n, \lambda, \Omega', \Omega'')$ such that the following properties hold:*

(i) $S(x, t) \subset \Omega''$ *for any $x \in \Omega'$, $0 \leq t \leq 2\rho$.*

(ii) *For all $\tau \in (0, 1)$ there exists $\beta = \beta(n, \lambda)$ such that $\tau S(x, t) \subset S(x, \tau t) \subset \tau^\beta S(x, t)$ for any $x \in \Omega'$, $0 \leq t \leq 2\rho$.*

(iii) **(Engulfing property)** *There exists a universal constant $\theta > 1$ such that, if $S(y, t) \cap S(x, t) \neq \emptyset$, then $S(y, t) \subset S(x, \theta t)$ for any $x, y \in \Omega'$, $0 \leq t \leq 2\rho/\theta$.*

(iv) $\cap_{0 < t \leq \rho} S(x, t) = \{x\}$.

*Proof.* Property (i) and (iv) are immediate by the strict convexity of $u$. The first inclusion in (ii) is just convexity, recall (2.19). To prove the second one it is enough to show the existence a universal constant $\sigma < 1$ such that

$$S(x, t/2) \subset \sigma S(x, t). \qquad (2.20)$$

Indeed, writing $\sigma = 2^{-\beta}$ we obtain the second inclusion in (ii) for $\tau = 1/2$. Iterating (recall (2.19))

$$S(x, t/2^k) \subset (2^{-k})^\beta S(x, t),$$

which implies the claim. But, after renormalization, the proof of (2.20) can be obtained exactly as in Step 1 of Theorem 2.1.

We are left to prove the engulfing property (iii). First notice that if we can prove that

$$\bar{y} \in S(x, t) \Rightarrow x \in S(\bar{y}, Kt)$$

for some universal $K$, then (iii) will follow with $\theta = K^2$. To show the above implication we can assume without loss of generality that $\nabla u(x) = 0$, thus

$$u(x) \leq u(y)$$
$$= u(y) + \nabla u(y) \cdot (x - y) + \nabla u(y) \cdot (y - x).$$

So our claim will follow if we can prove that

$$|\nabla u(y) \cdot (y - x)| \leq Kt \quad \forall y \in S(x, t).$$

To see notice that, up to reduce the size of $\rho$, we can assume that $S(x, 2t) \in \Omega$. Let $T$ the transformation which normalize $S(x, t)$ and let us consider the normalized solution:

$$v(z) = (\det T)^{2/n}[u(T^{-1}z) - 2t].$$

Then

$$\nabla u(y) \cdot (y - x) = \frac{1}{(\det T)^{2/n}} T^* \nabla v(Ty) \cdot (y - x)$$

$$= \frac{1}{(\det T)^{2/n}} \nabla v(Ty) \cdot (Ty - Tx).$$

Since $Ty$, $Tx$ belong to $B(0, n)$ and $(\det T)^{2/n} \approx 1/2t$ our claim is equivalent to

$$|\nabla v(z)| \leq \tilde{K} \quad \forall z \in T(S(x, t))$$

for some universal constant. By point (ii), $S(x, t) \subset 2^{-\beta} S(x, 2t)$ and hence

$$\text{dist}\left(T(S(x, t)), \partial T(S(x, 2t))\right) \geq \text{dist}\left(2^{-\beta} T(S(x, 2t)), \partial T(S(x, 2t))\right)$$
$$\geq 1/C(n, \lambda),$$

since $T(S(x, 2t))$ is normalized. Now by convexity and Lemma 2.6 for all $z \in T(S(x, t))$,

$$|\nabla v(z)| \leq \frac{|\inf_{T(S(x,2t))} v|}{\text{dist}\left(T(S(x, t)), \partial T(S(x, 2t))\right)} \leq \tilde{K}. \qquad \square$$

**Remark 2.13.** As shown by Forzani and Maldonado [65], the engulfing property of sections is actually equivalent to $C^{1,\alpha}$ regularity. Since the proof does not rely on compactness arguments one can also get the explicit dependence expression of the Hölder exponent by the structural constants. Here we briefly sketch the proof assuming that $u$ is differentiable. Let $y$ and $x$ be given, the smallest $t$ such that $u(x) \in S(y, t)$ is

$$\bar{t} = u(x) - u(y) - \nabla u(y) \cdot (x - y).$$

Thanks to the the engulfing property, $y \in S(x, \theta \bar{t})$ which implies

$$u(y) \leq u(x) + \nabla u(x) \cdot (y - x) + \theta\big(u(x) - u(y) - \nabla u(y) \cdot (x - y)\big).$$

Rearranging we get

$$(\nabla u(y) - \nabla u(x)) \cdot (y - x) \geq \frac{1+\theta}{\theta}\big(u(y) - u(x) - \nabla u(x) \cdot (y - x)\big).$$

Writing $y = x + sv$ with $|v| = 1$, we see that, for $s \leq \text{dist}(x, \partial\Omega)$, the function

$$\varphi(s) := u(x + sv) - u(x) - s\nabla u(x) \cdot v,$$

satisfies the differential inequality:

$$sf'(s) \geq \frac{1+\theta}{\theta} f(s).$$

Then $g(s) := f(s)s^{-(1+1/\theta)}$ is increasing, hence

$$\begin{aligned}
0 &\leq u(y) - u(x) + \nabla u(x) \cdot (y - x) \\
&= |x - y|^{1+1/\theta} g(|y - x|) \\
&\leq |x - y|^{1+1/\theta} g(\text{dist}(x, \partial\Omega)),
\end{aligned}$$

which implies that $u \in C^{1,1/\theta}$.

**Remark 2.14.** We remark here that sections with comparable height have comparable size. This means that there exists a constant $C$ depending on $n$, $\lambda$, $\Omega'$, $\Omega$ such that,

$$B(x, Ct) \subset S(x, t) \subset B(x, Ct^\beta) \qquad \forall x \in \Omega', \quad t \leq \rho \qquad (2.21)$$

where $\beta$ is as in Proposition 2.12. To see this, let $\rho$ be as in Proposition 2.12 with $\Omega'' = \Omega$. Since, by property (ii) of Proposition 2.12,

$$\frac{t}{\rho} S(x, \rho) \subset S(x, t) \subset \left(\frac{t}{\rho}\right)^\beta S(x, \rho) \qquad \text{for all } t \leq \rho,$$

it is enough to show that

$$B(x, r_1) \subset S(x, t) \subset B(x, r_2) \qquad (2.22)$$

for some $r_1$, $r_2$ which depends only on $\rho$ and on $\Omega$. To see this, let $E$ the John ellipsoid associated to $S(x, \rho)$ and let $\lambda_1 \leq \cdots \leq \lambda_n$ be the lengths of its semi-axis. From the inclusion $E \subset S(x, t) \subset nE$, we infer

$$2\lambda_n = \text{diam } E \leq \text{diam } S(x, t) \leq \text{diam } \Omega.$$

Since $|E| \gtrsim |S(x, \rho)| \gtrsim \rho^{n/2}$, this implies that $\lambda_1$ is bounded from below by a constant depending only on $\rho$ and $\Omega$. If $T$ is the transformation such that $T(E_\rho) = B_1$, then

$$\|T\| = \frac{1}{\lambda_1} \leq C(\rho, \Omega), \qquad \|T^{-1}\| = \lambda_n \leq C(\rho, \Omega).$$

So our claim will follow from

$$B(T(x), 1/C) \subset T(S(x, \rho)) \subset B(T(x), C)$$

for some universal $C$. The first inclusion follows by Aleksandrov maximum principle which implies that $\text{dist}(T(x), \partial T(S(x, \rho)) \geq 1/C$, while the second follows from the fact that, being $T(S(x, \rho))$ normalized,

$$T(S(x, \rho)) \subset B(0, n) \subset B(T(x), 3n),$$

since $T(x) \in B(0, n)$.

As we said, a particular nice feature of the above properties is that it is possible to prove classical theorems in Real Analysis using sections as they were balls, see [88]. In particular we quote the following which will play a crucial role.

**Lemma 2.15 (Vitali covering).** *Let $D$ be a compact set in $\Omega''$ and assume that to each $x \in D$ we associate a corresponding section $S(x, t_x) \in \Omega$. We can find a finite number of these sections $S(x_j, t_{x_j})$, $j = 1, \ldots, m$ such that*

$$D \subset \bigcup_{j=1}^{m} S(x_j, t_{x_j}), \qquad \text{with } S(x_j, t_{x_j}/\theta) \text{ disjoint.}$$

*Proof.* The proof follows as in the standard case. First, by compactness, we select a finite number of sections $\{S(x_i, t_{x_i}/\theta)\}_{i=1}^{N}$ which cover $D$. Among all these sections we choose one of maximal height, i.e. such that

$$t_{x_{i(1)}} = \max_{i=1,\ldots,N} t_{x_i},$$

and we discard all the ones which intersect $S(x_{i(1)}, t_{x_{i(1)}}/\theta)$. Among the remaining sections we choose again one of maximal height, $S(x_{i(2)}, t_{x_{i(2)}}/\theta)$, and we discard all the sections which intersect it. Continuing this process we obtain a finite number of disjoint sections $\{S(x_{i(j)}, t_{x_{i(j)}}/\theta)\}_{j=1}^{m}$ for which, thanks to the engulfing property,

$$D \subset \bigcup_{i=1}^{N} S(x_i, t_{x_i}/\theta) \subset \bigcup_{j=1}^{m} S(x_{i(j)}, t_{x_{i(j)}}). \qquad \square$$

Let $\Omega' \Subset \Omega'' \Subset \Omega''' \subset \Omega$ and $f \in L^1(\Omega'')$, we define the (localized) *maximal function*

$$M_{\Omega'',\Omega'''}[f](x) := \sup_{0<t<\rho} \fint_{S(x,t)} |f(y)|dy \qquad x \in \Omega'', \qquad (2.23)$$

where $\rho$ is such that for all $t \le \rho$ and $x \in \Omega''$, $S(x,t) \subset \Omega'''$, see Proposition 2.12 (i).

It is a classical theorem in Real Analysis that if the maximal function (done with respect to balls) $M_{\Omega'',\Omega'''}[f]$ is in $L^1$ then the function is in $L \log L(\Omega')$, meaning that

$$|f| \log_+ |f| \in L^1(\Omega'),$$

here $\log_+(t) = \max\{\log t, 0\}$.

The key estimates to prove the above result is valid in any space of homogeneous type, in particular it is also true for the maximal function constructed with sections. More precisely by [88, Chapter 1, Section 4, Theorem 2] and [88, Chapter 1, Section 8.14], the following key property holds: there exist universal constants $C'$, $C'' > 0$ such that, for any $\alpha \ge \alpha_0$,

$$\int_{\{f \ge \alpha\} \cap \Omega'} |f| \le C'\alpha \left|\{M_{\Omega',\Omega''}[f](x) \ge C''\alpha\}\right|. \qquad (2.24)$$

Here $\alpha_0$ is a sufficiently large constant which depends only on $\int_{\Omega'''} |f|$.

Here we give a sketch of the proof of the above inequality. First of all notice that exactly with the same proof as in the classical Euclidean setting the following *weak* $1 - 1$ estimate holds

$$\alpha \left|\{M_{\Omega'',\Omega'''}[f](x) > \alpha\}\right| \le C''' \int_{\Omega'''} |f|, \qquad (2.25)$$

where $C'''$ depends on the engulfing constant $\theta$ in Proposition 2.12. Moreover, since $M_{\Omega'',\Omega'''}[f]$ is easy seen to be lower semicontinuous, the set

$$\mathcal{M}_\alpha := \{M_{\Omega'',\Omega'''}[f](x) > \alpha\} \cap \Omega' \Subset \Omega''$$

is open. By Proposition 2.12 (i) there exists a $\sigma = \sigma(\Omega', \Omega'') > 0$ such that for all $x \in \Omega', t \le \sigma$,

$$S(y, 2\theta t) \subset \Omega'' \qquad \forall y \in S(x, 2t). \qquad (2.26)$$

By Remark 2.7,

$$|S(x,t)| \ge \frac{t^{n/2}}{\bar{C}} \qquad (2.27)$$

for some universal constant $\bar{C}$. Using (2.25) we choose $\alpha_0$ so big that

$$|\mathcal{M}_{\alpha_0}| \leq \left|\{M_{\Omega',\Omega''}[f](x) > \alpha_0\}\right| \leq \frac{\sigma^{n/2}}{\bar{C}}. \tag{2.28}$$

Notice that this condition fixes $\alpha_0$ in dependence only on $\int_{\Omega'} |f|$ and on the modulus of strict convexity of $u$ (and it is done just to be sure that all the sections we will consider are contained in $\Omega''$).

For $\alpha \geq \alpha_0$ and every $x \in \mathcal{M}_\alpha$ define

$$t(x) = \sup\{t : S(x,t) \subset \mathcal{M}_\alpha\}.$$

Thanks to (2.27) and (2.28), $t(x) \leq \sigma$, in particular (2.26) holds true for $t = t(x)$.

Let us fix $\varepsilon$ to be choosen and let us consider a covering of $\mathcal{M}_\alpha$ by sections $\{S(x, \varepsilon t(x))\}_{x \in \mathcal{M}_\alpha}$ and select a maximal disjoint subcollection $\{S(x_k, \varepsilon t(x_k))\}$.

We claim that (with the appropriate choice of $\varepsilon$)

$$\mathcal{M}_\alpha \subset \bigcup_k S(x_k, t(x_k)) \tag{2.29}$$

and

$$|\mathcal{M}_\alpha| \gtrsim \sum_k |S(x_k, t(x_k))|. \tag{2.30}$$

By maximality, for all $x \in \mathcal{M}_\alpha$,

$$S(x, \varepsilon t(x)) \cap S(x_k, \varepsilon t(x_k)) \neq \emptyset \tag{2.31}$$

for some $k$. Notice that $4\theta t(x_k) \geq t(x)$ if $\varepsilon < 1/2\theta$ since otherwise

$$S(x_k, 2t(x_k)) \cap S(x, t(x)/2\theta) \supset S(x_k, t(x_k)) \cap S(x, t(x)/2\theta) \neq \emptyset, \tag{2.32}$$

and $2t(x_k) \leq t(x)/2\theta$ implies, by the engulfing property and (2.32),

$$S(x_k, 2t(x_k)) \subset S(x, t(x)/2) \subset \mathcal{M}_\alpha,$$

a contradiction with the definition of $t(x)$. Since $4\theta\varepsilon t(x_k) \geq \varepsilon t(x)$, (2.31) implies

$$x \in S(x, 4\theta\varepsilon t(x_k)) \subset S(x_k, 4\theta^2\varepsilon t(x_k)).$$

Choosing $\varepsilon = 1/4\theta^2$ we obtain that $S(x_k, t(x_k))$ covers $\mathcal{M}_\alpha$, moreover, since $|S(x,t)| \approx t^{n/2}$, we have

$$|\mathcal{M}_\alpha| \geq \sum_k |S(x_k, \varepsilon t(x_k))| \gtrsim \sum_k |S(x_k, t(x_k))|,$$

and also (2.30) holds true.

We now prove (2.24). Since $S(x_k, 2t(x_k)) \cap (M_\alpha)^c \neq \emptyset$, we can find a $y \in S(x_k, 2t(x_k))$ such that (recall that, by (2.26), $S(y, 2t\theta(x_k)) \subset \Omega''$)

$$\fint_{S(x_k, 2t(x_k))} |f| \leq \theta^{n/2} \fint_{S(y, 2\theta t(x_k))} |f| \leq M_{\Omega'', \Omega'''}[f](y) \leq \theta^{n/2}\alpha.$$

Thus

$$|S(x_k, t(x_k))| \gtrsim |S(x_k, 2t(x_k))| \geq \frac{1}{\theta^{n/2}\alpha} \int_{S(x_k, t(x_k))} |f|.$$

Combining the above equation with (2.30) and (2.29), we obtain, for some universal constants $C'$, $C''$,

$$\int_{\{M_{\Omega'', \Omega'''}|f| \geq \alpha\} \cap \Omega'} |f| \leq C'\alpha \left|\{M_{\Omega'', \Omega'''}[f](x) \geq C''\alpha\}\right|.$$

Since, by Lebegue differentiation Theorem (which holds in spaces of homogeneous type, [88, Chapter 1]), for almost every $x \in \Omega''$

$$M_{\Omega'', \Omega'''}[f](x) \geq \alpha \quad \Longrightarrow \quad f(x) \geq \alpha,$$

we obtain (2.24).

## 2.3. Existence of smooth solutions to the Monge-Ampère equation

In this section we briefly recall the classical higher regularity theory for solutions to the Monge-Ampère equation. We will not give proofs of the results stated here but just sketch the main ideas behind them.

Existence of smooth solutions to the Monge Ampere equation (2.1) dates back to the work of Pogorelov. The way they are obtained (together with nice and useful estimates) is through the well-celebrated *method of continuity* which now we briefly describe (see [66, Chapter 17] for a more detailed exposition). Let us assume that we know how find a smooth (convex) solution $\bar{u}$ to

$$\begin{cases} \det D^2\bar{u} = \bar{f} & \text{in } \Omega \\ \bar{u} = 0 & \text{on } \partial\Omega \end{cases}$$

and that we would like to find a solution to

$$\begin{cases} \det D^2u = f & \text{in } \Omega \\ u = 0 & \text{on } \partial\Omega. \end{cases} \tag{2.33}$$

Let us define $f_t = (1 - t)\bar{f} + tf$, $t \in [0, 1]$, and let us consider the 1-parameter family of problems

$$\begin{cases} \det D^2 u_t = f_t & \text{in } \Omega \\ u_t = 0 & \text{on } \partial\Omega. \end{cases} \tag{2.34}$$

We would like to prove that the set of $t$ such that (2.34) is solvable is both open and closed, in this way we will clearly obtain a solution also to our original problem. More precisely let us assume that $f$, $\bar{f}$ are $C^{2,\alpha}(\overline{\Omega})$ and let us consider the set

$$\mathcal{C} = \{u : \overline{\Omega} \to \mathbb{R} \text{ convex functions of class } C^{4,\alpha}(\overline{\Omega}), u = 0 \text{ on } \partial\Omega\}.$$

Notice that $\mathcal{C}$ is non-trivial if and only if $\Omega$ is a $C^{4,\alpha}$ convex set. Consider the non-linear map

$$\mathcal{F}: \mathcal{C} \times [0, 1] \longrightarrow C^{2,\alpha}(\overline{\Omega})$$
$$(u, t) \mapsto \det D^2 u - f_t.$$

We would like to show that

$$\mathcal{T} = \{t \in [0, 1] : \text{ there exists a } u_t \in \mathcal{C} \text{ such that } \mathcal{F}(u_t, t) = 0\},$$

is both open and closed. Openess follows from the Implicit Function Theorem in Banach spaces (see [66, Theorem 17.6]). Indeed, the Frechèt differential of $\mathcal{F}$ with respect to $u$ is given by the linearized Monge-Ampère operator (see also Chapter 5):

$$\mathcal{L}[h] := D_u\mathcal{F}(u, t)[h] = M_{ij}(D^2 u)D_{ij}h, \qquad h = 0 \text{ on } \partial\Omega. \tag{2.35}$$

Here $M_{ij}(D^2 u)$ is the cofactor matrix of $D^2 u$.[5] Notice that if $u$ is bounded in $C^2$ and $f$ is bounded from below by $\lambda$, then the smallest eigenvalue of $D^2 u$ is also bounded from below, in this way the (2.35) operator becomes uniformly elliptic with $C^{2,\alpha}$ coefficients. Classical Schauder theory gives then the invertibility of $D_u\mathcal{F}(u, t)$.

---

[5] Given an invertible matrix $A$ the cofactor matrix $M(A)$ is the one which satisfies

$$M(A)^*A = (\det A)\,\text{Id}.$$

For non-invertible matrices the definition is extended by continuity. Notice that the above equation implies

$$\frac{\partial \det A}{\partial A_{ij}} = M_{ji}(A).$$

The task is now to prove closedness of $\mathcal{T}$. This is done through *a-priori estimates* both at the interior and at the boundary. As already said the Monge-Ampère equation becomes uniformly elliptic (meaning that the linearized equation is uniformly elliptic) on uniformly convex functions, thus the main task is to establish an a-priori bound on the $C^2$ norm of $u$ in $\overline{\Omega}$, since then the lower bound on the determinant will imply that the smallest eigenvalue of $D^2u$ is bounded away from 0. Once this is done, classical Schauder theory gives higher regularity.

For what concerns interior estimates we have the following classical theorem due to Pogorelov. The rate of degeneracy of the estimates can be found in [28].

**Theorem 2.16 (Pogorelov).** *Let $u$ be a $C^4(\Omega)$ solution to (2.33). Assume that $B_1 \subset \Omega \subset B_n$, and that $\lambda \leq f \leq 1/\lambda$. Then there exist a constant $C$ depending only on $n$, $\lambda$ and $\|f\|_{C^2}$ and an universal exponent $\tau$ such that*

$$\left( \operatorname{dist}(x, \partial\Omega) \right)^\tau |D^2u(x)| \leq C \qquad \forall\, x \in \Omega. \tag{2.36}$$

The necessity of working with normalized domains and solutions is given by the family of functions

$$u_\varepsilon(x, y) = \frac{\varepsilon x^2}{2} + \frac{y^2}{2\varepsilon} - 1$$

which solves $\det D^2 u_\varepsilon = 1$ on $\{u_\varepsilon \leq 0\}$.

To obtain estimates up to the boundary we also need to impose some assumption on the domain, [66, Theorem 17.20]:

**Theorem 2.17.** *Let $\Omega$ be a uniformly convex $C^3$ domain [6] and let $u$ be a solution (2.33) with $f \in C^2(\overline{\Omega})$ and $\lambda \leq f \leq 1/\lambda$. Then there exists a constant $C$ depending on $\Omega$, $\lambda$, $\|f\|_{C^2(\overline{\Omega})}$ such that*

$$\|D^2u\|_{C^0(\overline{\Omega})} \leq C.$$

Combining the above Theorems one gets the following existence result:

**Theorem 2.18.** *Let $\Omega$ be a uniformly convex $C^{4,\alpha}$ domain. Then for all $f \in C^{2,\alpha}(\overline{\Omega})$, $\lambda \leq f \leq 1/\lambda$ there exists a (unique) $C^{4,\alpha}(\overline{\Omega})$ solution to (2.33).*

---

[6] We say that a domain is uniformly convex if its second fundamental form is (as a symmetric tensor) uniformly bounded from below. This is equivalent to require the existence of a uniformly convex function $\phi$ such that $\Omega = \{\phi < 0\}$.

Notice that the above Theorem actually gives also existence of Aleksandrov solutions to (2.33) with $f$ merely bounded away from zero and infinity and $\Omega$ not strictly convex. Indeed one can approximate $f$ and $\Omega$ with a sequence $f_k$ and $\Omega_k$ satisfying the hypothesis of the above Theorem, find a sequence of solutions $u_k$ and apply the results of the previous section (see Lemma 2.10 for instance) to show that the $u_k$ converges to a function $u$ solving (2.33). For a direct approach to the existence of Aleksandrov solutions, see [67].

Theorem 2.16 needs to assume $f \in C^2$ to obtain uniform bounds on the $C^2$ norm of $u$, actually as shown by Caffarelli [19] much less is needed.

**Theorem 2.19.** *Let $u$ be a solution of* (2.33) *and assume that $f \in C^{\alpha}_{\mathrm{loc}}(\Omega)$, then $u \in C^{2,\alpha}_{\mathrm{loc}}(\Omega)$.*

Later, in [69], this result has been improved by showing that if $f$ is Dini-continuous then $u$ is $C^2$. The examples in [93] imply that this result is essentially optimal.

The proof of the above theorem is based on showing that under the assumption that $f$ is almost a constant, say 1, then $u$ is very close to the solution of (2.33) with right hand side 1. Since this latter function has interior a priori estimates we can prove by interpolation that the $C^2$ norm of $u$ remains bounded (see the proof of Theorem 6.11). With this line of reasoning one can also prove the following Theorem, also due to Caffarelli [19]

**Theorem 2.20.** *Let $u$ be a solution of* (2.33), *then for every $p > 1$ there exists a $\delta$ such that if $\|f - 1\|_\infty \leq \delta$, then $u \in W^{2,p}_{\mathrm{loc}}$.*

We will give a simple proof of this Theorem in Chapter 3.

**Example 2.21.** Let us stress here that to obtain $u \in W^{2,p}$ the "pinching" $\|f - 1\|_\infty \leq \delta$ is necessary. Indeed in [93] Wang shows that the family of convex functions

$$u_\alpha(x, y) = \begin{cases} x^\alpha + \frac{\alpha^2 - 1}{\alpha(\alpha-2)} y^2 x^{2-\alpha} & \text{for } |y| \leq |x|^{\alpha-1} \\[2ex] \frac{1}{2\alpha} x^2 y^{(\alpha-2)/(\alpha-1)} + \frac{4\alpha-5}{2(\alpha-2)} y^{\alpha/(\alpha-1)} & \text{for } |y| \geq |x|^{\alpha-1} \end{cases}$$

are strictly convex solution to (2.33) with $\lambda(\alpha) \leq f \leq 1/\lambda(\alpha)$ but $u_\alpha \notin W^{2,p}$ for $p > \alpha/(\alpha - 2)$.

In spite of the above example in Chapter 3 we will prove that solution of (2.33) are always of class $W^{2,1}$. Notice that, due to the lack of regularity of $f$, a "perturbative" approach cannot work in this general case.

# Chapter 3
# Sobolev regularity of solutions to the Monge Ampère equation

In this Chapter we prove the $W^{2,1}$ regularity of solutions of (2.1). This has been first shown in [40] in collaboration with Alessio Figalli, where actually the following higher integrability result was proved

**Theorem 3.1.** *Let* $\Omega \subset \mathbb{R}^n$ *be a bounded convex domain, and let* $u :$ $\overline{\Omega} \to \mathbb{R}$ *be an Aleksandrov solution of* (2.1) *with* $0 < \lambda \leq f \leq 1/\lambda$. *Then* $u \in W_{\text{loc}}^{2,1}(\Omega)$ *and for any* $\Omega' \Subset \Omega$ *and* $k \in \mathbb{N} \cup \{0\}$, *there exists a constant* $C = C(k, n, \lambda, \Omega, \Omega') > 0$ *such that*

$$\int_{\Omega'} \|D^2 u\| \log_+^k \left(\|D^2 u\|\right) \leq C. \tag{3.1}$$

Later, in collaboration with Alessio Figalli and Ovidiu Savin [44] and independently by Thomas Schmidt [86], the $L \log^k L$ integrability has been improved to $L^{1+\varepsilon}$ where $\varepsilon = \varepsilon(n, \lambda)$. Notice that in view of Wang counterexample, Example 2.21, the result is optimal.

**Theorem 3.2.** *Let* $\Omega \subset \mathbb{R}^n$ *be a bounded convex domain, and let* $u :$ $\overline{\Omega} \to \mathbb{R}$ *be an Aleksandrov solution of* (2.1) *with* $0 < \lambda \leq f \leq 1/\lambda$. *Then there exists a* $\gamma_0 = \gamma_0(n, \lambda) > 1$ *such that, for any* $\Omega' \Subset \Omega$ *there exists* $C = C(n, \lambda, \Omega, \Omega') > 0$ *for which the following estimate holds*

$$\int_{\Omega'} \|D^2 u\|^{\gamma_0} \leq C. \tag{3.2}$$

As a corollary we obtain the following Sobolev regularity result for optimal transport maps

**Corollary 3.3.** *Let* $\Omega_1, \Omega_2 \subset \mathbb{R}^n$ *be two bounded domains, and* $\rho_1, \rho_2$ *two probability densities such that* $0 < \lambda \leq \rho_1, \rho_2 \leq 1/\lambda$ *inside* $\Omega_1$ *and* $\Omega_2$ *respectively. Let* $T = \nabla u : \Omega_1 \to \Omega_2$ *be the (unique) optimal transport map for the quadratic cost sending* $\rho_1$ *onto* $\rho_2$, *and assume that* $\Omega_2$ *is convex. Then* $T$ *is locally in* $W^{1,1}$, *more precisely* $T \in W_{\text{loc}}^{1,\gamma_0}(\Omega_1)$ *with* $\gamma_0 = \gamma_0(n, \lambda) > 1$.

As we saw at the end of Section 2.2, many theorems in Real Analysis hold replacing balls with sections. A key tool in the proof of Theorem 3.1 will be the maximal operator of $D^2u$, the hessian of a $C^2$ solution of (2.1), namely if $\Omega'' \Subset \Omega''' \subset \Omega$ we define

$$M_{\Omega'',\Omega'''}(x) := \sup_{0<t<\rho} \fint_{S(x,t)} \|D^2u(y)\| \, dy \qquad x \in \Omega'' \tag{3.3}$$

where $\|D^2u(y)\|$ denotes the operator norm of the matrix $D^2u(y)$, and $\rho$ is such that $S(x,t) \subset \Omega'''$ for all $x \in \Omega''$, $t \le \rho$, see Proposition 2.12.

Estimate (2.24) then becomes, for $\Omega' \Subset \Omega'' \Subset \Omega'''$

$$\int_{\{|D^2u\| \ge \alpha\} \cap \Omega'} \|D^2u\| \le C'\alpha \left| \{|M_{\Omega',\Omega''}(x) \ge C''\alpha\} \cap \Omega'' \right| \qquad \forall \alpha \ge \alpha_0. \tag{3.4}$$

Here $C'$ and $C''$ are universal constants and $\alpha_0$ depends only on $\int_{\Omega'''} \|D^2u\|$.

The strategy of the proof of Theorem 3.1 will be then to prove the following estimate for $C^2$ solutions of (2.1)

$$\left| \{|M_{\Omega',\Omega''}(x) \gtrsim \alpha\} \cap \Omega'' \right| \lesssim \left| \{\|D^2u\|(x) \ge \alpha\} \cap \Omega' \right|,$$

where all the involved constants are universal. This will lead to an a-priori estimate for the $L \log^k L$ norms of $D^2u$. Since any solution of (2.1) can be approximated by smooth solutions (regularizing the data and applying Theorem 2.18) and all bounds depend only on $n$ and $\lambda$, this will prove Theorem 3.1. A more careful analysis will then lead to Theorem 3.2.

## 3.1. Proof of Theorem 3.1

As we said it suffices to prove (3.1) when $u \in C^2(\Omega)$.

Let us remark that the proof of (3.1) for $k = 0$ is elementary: indeed, this follows from $\|D^2u\| \le \Delta u$ (since $u$ is convex) and a universal interior bound for the gradient of $u$ (see for instance the proof of point (iii) of Proposition 2.12 or (3.11)-(3.12) below).

Hence, performing an induction on $k$ and using a standard covering argument, briefly sketched below, it suffices to prove the following result (recall the notation (2.19) for the dilation of a section):

**Theorem 3.4.** Let $\mathcal{U} \subset \mathbb{R}^n$ be a normalized convex set, and let $u : \mathcal{U} \to \mathbb{R}$ a $C^2$ convex solution of

$$\begin{cases} 0 < \lambda \le \det D^2u \le 1/\lambda & in \, \mathcal{U}, \\ u = 0 & on \, \partial\mathcal{U}. \end{cases} \tag{3.5}$$

*Then for any $k \in \mathbb{N} \cup \{0\}$ there exists a constant $C = C(k, n, \lambda)$ such that*

$$\int_{\mathcal{U}/2} \|D^2 u\| \log_+^{k+1} \left( \|D^2 u\| \right) \leq C \int_{3\mathcal{U}/4} \|D^2 u\| \log_+^k \left( \|D^2 u\| \right).$$

*Proof of* (3.1) *using Theorem* 3.4. We want to show that, if (3.1) holds for some $k \in \mathbb{N} \cup \{0\}$, then it also holds for $k + 1$. Since the following argument is standard, we just give a sketch of the proof.

Given $\Omega' \subset\subset \Omega$, fix $\rho$ as in Proposition 2.12 with $\Omega'' = \Omega$, and consider the covering of $\Omega'$ given by $\{S(x, \rho)\}_{x \in \Omega'}$. By Remark 2.14 (see (2.22)) all sections $\{S(x, \rho)\}_{x \in \Omega'}$ have comparable shapes, there exist positive constants $r_1, r_2$, depending only on $n, \lambda, , \rho, \Omega$, such that

$$B(x, r_1) \subset S(x, \rho) \subset B(x, r_2) \qquad \forall x \in \Omega'. \tag{3.6}$$

This implies that one can cover $\Omega'$ with finitely many such sections $\{S_i\}_{i=1,\ldots,N}$ (the number $N$ depending only on $r_1, r_2, \Omega'$), and moreover the affine transformations $T_i$ normalizing them (which we assume to be symmetric) satisfy the following bounds (which follow easily from (3.6) and the inclusion $B(0, 1) \subset T_i(S_i) \subset B(0, n)$):

$$\|T_i\| \leq \frac{n}{r_1}, \qquad \det T_i \geq \frac{1}{r_2^n}.$$

Hence, we can define $v_i$ as in (2.11) with $T = T_i$ and $t = \rho$, and apply Theorem 3.4 to each of them: by using the inductive hypothesis we have

$$\int_{T_i(S_i)/2} \|D^2 v_i\| \log_+^{k+1} \left( \|D^2 v_i\| \right) \leq C(k, n, \lambda).$$

Changing variables back and summing over $i$, we get

$$\int_{\Omega'} \|D^2 u\| \log_+^k \left( \|D^2 u\| \right)$$

$$\leq C(k, n, \lambda) \sum_{i=1}^N \frac{\|T_i\|^2}{(\det T_i)^{1+2/n}} \log_+^k \left( \frac{\|T_i\|^2}{(\det T_i)^{2/n}} \right).$$

Recalling that we have uniform bounds on $N$ and on $T_i$, this concludes the proof. $\qquad \square$

We now focus on the proof of Theorem 3.4. We begin by showing that the average of $\|D^2 u\|$ over a section is controlled by the size of the "normalizing affine transformation".

**Lemma 3.5.** *Let $u$ solve* (3.5)*, fix $x \in \mathcal{U}/2$, and let $t > 0$ be such that $S(x, 2t) \subset 3\mathcal{U}/4$. Let $T$ be the (symmetric) affine map which normalizes $S(x, t)$. Then there exists a positive universal constant $C_1$ such that*

$$\frac{\|T\|^2}{(\det T)^{2/n}} \geq C_1 \fint_{S(x,t)} \|D^2u\|. \tag{3.7}$$

*Proof.* Let us consider $v : T(S(x, 2t)) \to \mathbb{R}$, with $v$ is defined as in (2.11). We notice that

$$D^2v(z) = (\det T)^{2/n} \left[ (T^{-1})^* D^2u(T^{-1}z)T^{-1} \right], \tag{3.8}$$

and

$$\begin{cases} \lambda \leq \det D^2v \leq 1/\lambda & \text{in } T(S(x, 2t)), \\ v = t(\det T)^{2/n} & \text{on } \partial\big(T(S(x, 2t))\big). \end{cases} \tag{3.9}$$

Although the convex set $T(S(x, 2t))$ is not renormalized in the sense defined before, it is almost so: indeed, since $T$ normalizes $S(x, t)$ we have $Tz \in B(0, n)$ for any $z \in S(x, t)$. Recalling that $2S(x, t)$ denotes the dilation of $S(x, t)$ with respect to $x$ (see (2.19)), we get

$$T\left(x + \frac{y - x}{2}\right) \in B(0, n) \qquad \forall\, y \in 2S(x, t),$$

which is equivalent to

$$Ty + Tx \in B(0, 2n) \qquad \forall\, y \in 2S(x, t).$$

Since $Tx \in B(0, n)$ this implies that $T(2S(x, t)) \subset B(0, 3n)$, which together with the fact that $S(x, t) \subset S(x, 2t) \subset 2S(x, t)$ (by convexity of $u$) gives

$$B(0, 1) \subset T(S(x, 2t)) \subset B(0, 3n). \tag{3.10}$$

Hence, it follows from (3.9), $\det T^{2/n}t \approx 1$ and Lemma 2.6 that

$$\underset{T(S(x,2t))}{\text{osc}} v = \left| \underset{T(S(x,2t))}{\inf} (v - t(\det T)^{2/n}) \right| \leq c', \tag{3.11}$$

with $c'$ universal.

Since $v$ is convex, by (3.11), and Proposition 2.12(ii), we also get (cp. the proof of Propositon 2.12 (iii) )

$$\begin{aligned} \underset{T(S(x,t))}{\sup} |\nabla v| &\leq \underset{\beta T(S(x,2t))}{\sup} |\nabla v| \\ &\leq \frac{\text{osc}_{T(S(x,2t))}\, v}{\text{dist}\big(2^{-\beta}T(S(x, 2t)),\, \partial\big(T(S(x, 2t))\big)\big)} \leq c'' \end{aligned} \tag{3.12}$$

for some universal constant $c''$. Moreover, since $T(S(x,t))$ is a normalized convex set, it holds

$$|B_1| \leq |T(S(x,t))| = \det T |S(x,t)|, \quad \mathcal{H}^{n-1}(\partial T(S(x,t))) \leq c(n), \quad (3.13)$$

where $c(n)$ is a dimensional constant. Finally, using again the convexity of $v$, the estimate

$$\|D^2 v(y)\| \leq \Delta v(y) \qquad (3.14)$$

holds (recall that $\|D^2 u(y)\|$ denotes the operator norm of $D^2 u(y)$).

Hence, by (3.8), (3.13), (3.14) and (3.12) and since $T$ is symmetric, we get

$$\fint_{S(x,t)} \|D^2 u(y)\| \, dy = \frac{1}{(\det T)^{2/n}} \fint_{S(x,t)} \|T^* D^2 v(Ty) T\| \, dy$$

$$\leq \frac{\|T\|^2}{(\det T)^{2/n}} \frac{1}{\det T |S(x,t)|} \int_{T(S(x,t))} \|D^2 v(z)\| \, dz$$

$$\leq \frac{\|T\|^2}{(\det T)^{2/n} \omega_n} \int_{T(S(x,t))} \Delta v(z) \, dz$$

$$= \frac{\|T\|^2}{(\det T)^{2/n} \omega_n} \int_{T(\partial S(x,t))} \nabla v(z) \cdot \nu \, d\mathcal{H}^{n-1}(z)$$

$$\leq \frac{c(n)\|T\|^2}{(\det T)^{2/n} \omega_n} \sup_{T(S(x,t))} |\nabla v|$$

$$\leq c'' \frac{c(n)\|T\|^2}{(\det T)^{2/n} \omega_n},$$

which concludes the proof of (3.7). □

We now show that, in every section, we can find a uniform fraction of points where the norm of the Hessian controls the size of the "normalizing affine transformation".

**Lemma 3.6.** *Let $u$ solve* (3.5), *fix $x \in \mathcal{U}/2$, and let $t > 0$ be such that $S(x, 2t) \subset 3\mathcal{U}/4$. Let $T$ be the (symmetric) affine map which normalizes $S(x,t)$ and let $\theta$ be the engulfing constant of Proposition 2.12 (iii). Then there exist universal positive constants $C_2$, $C_3$, and a Borel set $A(x,t) \subset S(x, t/\theta)$, such that*

$$\frac{|A(x,t) \cap S(x, t/\theta)|}{|S(x,t)|} \geq C_2, \qquad (3.15)$$

*and*

$$\|D^2 u(y)\| \geq C_3 \frac{\|T\|^2}{(\det T)^{2/n}} \qquad \forall y \in A(x,t). \qquad (3.16)$$

*Proof.* We divide the proof in two steps.

*Step one: Let $v$ be a normalized solution in $Z$ (see* (2.12)*). Then there exist universal constants $c'$, $c'' > 0$, and a Borel set $E \subset Z$, such that $|E| \geq c'|Z|$, and $D^2 v(x) \geq c'' \mathrm{Id}$ for every $x \in E$.*

To see this, let us consider the paraboloid $p(x) := c_1(|x|^2/n^2 - 1)/2$, with $c_1$ as in (2.13) (observe that, since $Z \subset B(0, n)$, $p \leq 0$ inside $Z$). Then

$$|\inf_\Omega(v - p)| \geq \frac{c_1}{2}.$$

Set $w := v - p$, and let $\Gamma_w : Z \to \mathbb{R}$ be a convex envelope of $w$ in $Z$, that is

$$\Gamma_w(y) := \sup\{\ell(y) : \ell \leq w \text{ in } Z, \ell \leq 0 \text{ on } \partial Z, \ell \text{ affine}\}.$$

It is well-known that $\Gamma_w$ is $C^{1,1}(Z)$ (see [42] for instance), and that $\det D^2\Gamma_w = 0$ outside the set $\{\Gamma_w = w\} \subset Z$ (see (4.6)). Hence, by Aleksandrov maximum principle, Theorem 2.4, and from the fact that

$$0 \leq D^2\Gamma_w \leq D^2 w \leq D^2 v \qquad \text{a.e. on } \{\Gamma_w = w\}$$

(in the sense of non-negative symmetric matrices), we get

$$\left(\frac{c_1}{2}\right)^n \leq \left|\inf_Z w\right|^n = \left|\inf_Z \Gamma_w\right|^n \leq C(n) \int_{\{\Gamma_w=w\}} \det D^2\Gamma_w$$

$$\leq C(n) \int_{\{\Gamma_w=w\}} \det D^2 v \leq C(n)\left|\{\Gamma_w = w\}\right|/\lambda.$$

This provides a universal lower bound on the measure of $E := \{\Gamma_w = w\}$. Moreover, since $D^2 w \geq 0$ on $E$, we obtain

$$D^2 v \geq \frac{c_1}{n^2}\mathrm{Id} \qquad \text{on } E,$$

proving the claim.

*Step two: Proof of the Lemma.* Let $S(x, t/\theta) \subset S(x, t)$ and $T_\theta$ be the (symmetric) affine transformation which normalizes $S(x, t/\theta)$. Define $v$ as in (2.11) (using $T_\theta$). Since $v$ is a normalized solution, we can apply the previous step to find a set $E \subset Z := T_\theta(S(x, t/\theta))$ such that $|E| \geq c'|Z|$, and $D^2 v \geq c''\mathrm{Id}$ on $E$. We define $A(x, t) := T_\theta^{-1}(E) \subset S(x, t/\theta)$.

To prove (3.15) we observe that, since $S(x, t/\theta) \approx (t/\theta)^n \approx |S(x, t)|$ (recall that $\theta$ is universal), we have

$$\frac{|A(x, t) \cap S(x, t/\theta)|}{|S(x, t)|} \geq c''' \frac{|A(x, t) \cap S(x, t/\theta)|}{|S(x, t)|} = c''' \frac{|E \cap Z|}{|Z|} \geq c'''c'.$$

Moreover, since on $A(x, t)$

$$D^2 u(y) = \frac{1}{(\det T_\theta)^{2/n}} T_\theta^* D^2 v(Ty) T_\theta \geq \frac{c''}{(\det T)^{2/n}} T_\theta^* T_\theta,$$

using (2.10) we get

$$\|D^2 u(y)\| \geq \frac{c'' \|T_\theta^* T_\theta\|}{(\det T_\theta)^{2/n}} = \frac{c'' \|T_\theta^*\|^2}{(\det T_\theta)^{2/n}} \qquad \forall y \in A(x, t).$$

To conclude the proof of (3.16) we need to show that, if $T$ is the affine transformation which normalizes $T$, then

$$\det T_\theta \leq C \det T \quad \text{and} \quad \|T_\theta\| \geq \|T\|/C$$

for some universal constant $C$. The first inequality trivially follows from $1/\det T_\theta \approx |S(x, t/\theta)| \approx (t/\theta)^{n/2}$ and $1/\det T \approx |S(x, t)| \approx t^{n/2}$. To prove the second, recall that by Proposition 2.12 (ii)

$$S(x, t)/\theta \subset S(x, t/\theta) \subset (1/\theta)^{-\beta} S(x, t).$$

Hence arguing as in the proof of Lemma 3.5, for some universal constant $C'(n, \theta)$

$$T_\theta T^{-1}(B_1) \subset T_\theta T^{-1}(T(S(x, t))) \subset T_\theta(\theta S(x, t/\theta)) \subset C' B_n,$$

from which we deduce

$$\|T_\theta T^{-1}\| \leq C.$$

Thus

$$\|T_\theta\| = \|T_\theta T^{-1} T\| \leq C \|T\|. \qquad \square$$

Combining the two previous lemmas, we obtain that in every section we can find a uniform fraction of points where the norm of the Hessian controls its average over the section:

$$\|D^2 u(y)\| \geq C_1 C_3 \fint_{S(x,t)} \|D^2 u\| \quad \forall y \in A(x, t) \subset S(x, t/\theta). \quad (3.17)$$

As we will show below, Theorem 3.4 is a direct consequence of this fact and a covering argument.

To simplify the notation, we use $M(x)$ to denote $M_{\mathcal{U}/2, 3\mathcal{U}/4}(x)$ (see (3.3)).

**Lemma 3.7.** *Let u solve* (3.5). *Then there exist two universal positive constants $C_4$ and $C_5$ such that*

$$|\{x \in \mathcal{U}/2 : \ M(x) > \gamma\}| \le C_4 |\{x \in 3\mathcal{U}/4 : \ \|D^2u(x)\| \ge C_5\gamma\}| \quad (3.18)$$

*for every $\gamma > 0$.*

*Proof.* By the definition of $M$, we clearly have

$$\{x \in \mathcal{U}/2 : \ M(x) \ge \gamma\} \subset E := \left\{x \in \mathcal{U}/2 : \ \fint_{S(x,t_x)} |D^2u| \ge \frac{\gamma}{2}\right.$$

$$\left. \text{for some } t_x \in (0, \rho)\right\}.$$

By Lemma 2.15, we can find a finite numbers of points $x_k \in E$ such that $\{S(x_k, t_{x_k})\}$ covers $E$ and $\{S(x_k, t_{x_k}/\theta)\}$ are disjoints. Since $x_k \in E$, by (3.17) we deduce that

$$\|D^2u(y)\| \ge C_1 C_3 \fint_{S(x_k,t_{x_k})} \|D^2u\|$$

$$\ge \frac{C_1 C_3 \gamma}{2} \quad \forall y \in A(x_k, t_{x_k}) \subset S(x_k, t_{x_k}/\theta). \quad (3.19)$$

Hence, applying (3.15), (3.19), we get

$$|\{x \in \mathcal{U}/2 : \ M(x) \ge \gamma\}| \le \sum_{k \in \mathbb{N}} |S(x_k, t_{x_k})|$$

$$\le \frac{1}{C_2} \sum_{k \in \mathbb{N}} |A(x_k, t_{x_k}) \cap S(x_k, t_{x_k}/\theta)|$$

$$\le \frac{1}{C_2} \sum_{k \in \mathbb{N}} |S(x_k, t_{x_k}/\theta) \cap \{x \in 3\mathcal{U}/4 : \ \|D^2u(x)\| \ge C_1 C_3 \gamma/2\}|$$

$$= \frac{1}{C_2} \left| \bigcup_k S(x_k, t_k/\theta) \cap \{x \in 3\mathcal{U}/4 : \ \|D^2u(x)\| \ge C_1 C_3 \gamma/2\} \right|$$

$$\le \frac{1}{C_2} |\{x \in 3\mathcal{U}/4 : \ \|D^2u(x)\| \ge C_1 C_3 \gamma/2\}|,$$

proving the result. $\qquad\qquad\qquad\qquad\qquad\qquad\qquad\qquad\qquad\qquad\qquad\square$

*Proof of Theorem* 3.4. Combining (3.4) and (3.18), we obtain the existence of two positive universal constants $c'$, $c''$ such that

$$\int_{\mathcal{U}/2 \cap \{\|D^2u\| \ge \gamma\}} \|D^2u\| \le c'\gamma |\{x \in 3\mathcal{U}/4 : \ \|D^2u(x)\| \ge c''\gamma\}| \quad \forall \gamma \ge \bar{c},$$

$$(3.20)$$

with $\bar{c}$ depending only on $\fint_{3\mathcal{U}/4}\|D^2u\|$ and $\rho$. Observe that, since $u$ is normalized, both $\fint_{3\mathcal{U}/4}\|D^2u\|$ and $\rho$ are universal (see the discussion at the beginning of this section), so $\bar{c}$ is universal as well. In addition, we can assume without loss of generality that $\bar{c} \geq 1$. So

$$\int_{\mathcal{U}/2}\|D^2u\|\log_+^{k+1}(\|D^2u\|)$$

$$\leq \log_+^k(\bar{c})\int_{\mathcal{U}/2\cap\{\|D^2u\|\leq\bar{c}\}}\|D^2u\| + \int_{\mathcal{U}/2\cap\{\|D^2u\|\geq\bar{c}\}}\|D^2u\|\log_+^{k+1}(\|D^2u\|)$$

$$\leq C(n)\bar{c}\log_+^k\bar{c} + \int_{\mathcal{U}/2\cap\{\|D^2u\|\geq\bar{c}\}}\|D^2u\|\log^{k+1}\|D^2u\|.$$

Hence, to prove the result, it suffices to control the last term in the right hand side. We observe that such a term can be rewritten as

$$2(k+1)\int_{\mathcal{U}/2\cap\{\|D^2u\|\geq\bar{c}\}}\|D^2u\|\int_1^{\|D^2u\|}\frac{\log^k(\gamma)}{\gamma}\,d\gamma,$$

which is bounded by

$$C' + 2(k+1)\int_{\mathcal{U}/2\cap\{\|D^2u\|\geq\bar{c}\}}\|D^2u\|\int_{\bar{c}}^{\|D^2u\|}\frac{\log^k(\gamma)}{\gamma}\,d\gamma,$$

with $C' = C'(k,\bar{c})$. Now, by Fubini, the second term is equal to

$$2(k+1)\int_{\bar{c}}^{\infty}\frac{\log^k(\gamma)}{\gamma}\left(\int_{\mathcal{U}/2\cap\{\|D^2u\|\geq\gamma\}}\|D^2u\|\right)d\gamma,$$

which by (3.20) is controlled by

$$2(k+1)c'\int_{\bar{c}}^{\infty}\log^k(\gamma)\,|\{x\in 3\mathcal{U}/4:\ \|D^2u(x)\|\geq c''\gamma\}|\,d\gamma.$$

By the layer-cake representation formula and, since $\bar{c} \geq 1$, this last term is bounded by

$$C''\int_{3\mathcal{U}/4}\|D^2u\|\log_+^k(\|D^2u\|)$$

for some $C'' = C''(k,c',c'')$, concluding the proof. $\qquad\square$

## 3.2. Proof of Theorem 3.2

In this Section ,we show how to improve the $L \log^k L$ integrability of the previous section, to a $L^{\gamma_0}$ integrability for some $\gamma_0 = \gamma_0(\lambda, n) > 1$. We will follow the presentation in [44]. As we already mentioned a different proof was achieved independently by Thomas Schmidt in [86]. He uses some abstract harmonic analysis on sections (see the comments after Proposition 2.12) to show how from the maximal inequality (3.18) (which can be thought as a "reverse" $L^1$-weak $L^1$ Hölder inequality) the higher integrability follows from a Gehring type lemma.

As in the proof of Theorem 3.1, to prove Theorem 3.2 it will be enough to prove the following "normalized" statement.

**Theorem 3.8.** Let $u : \overline{\mathcal{U}} \to \mathbb{R}$, be a $C^2$ solution of (2.1) with $\lambda \leq f \leq 1/\lambda$ and let us assume that $B_1 \subset \mathcal{U} \subset B_n$, then there exist universal constants $C$ and $\varepsilon_0 > 0$ such that

$$\int_{B_{1/2}} \|D^2 u\|^{1+\varepsilon_0} \leq C. \tag{3.21}$$

Theorem 3.8 follows by slightly modifying the strategy in the previous section: we use a covering lemma that is better localized (see Lemma 3.10) to obtain a geometric decay of the "truncated" $L^1$ energy for $\|D^2 u\|$ (see Lemma 3.12).

We also give a second proof of Theorem 3.8 based on the following observation: in view of Theorem 3.1 the $L^1$ norm of $\|D^2 u\|$ decays on sets of small measure:

$$|\{\|D^2 u\| \geq M\}| \leq \frac{C}{M \log M},$$

for an appropriate universal constant $C > 0$ and for any $M$ large. In particular, choosing first $M$ sufficiently large and then taking $\varepsilon > 0$ small enough, we deduce (a localized version of) the bound

$$|\{\|D^2 u\| \geq M\}| \leq \frac{1}{M^{1+\varepsilon}}|\{\|D^2 u\| \geq 1\}|.$$

Applying this estimate at all scales (together with Lemma 2.15) leads to the local $W^{2,1+\varepsilon}$ integrability for $\|D^2 u\|$.

In the proof of Theorem 3.1 the following quantity played a distinguished role: if $S(x, t)$ is a section of $u$ and $T$ is the (symmetric) affine transformation that normalize it, that is such that

$$B_1 \subset T(S(x, t)) \subset B_n,$$

we define the *normalized size* of $S(x, t)$ to be

$$\alpha(S(x, t)) = \frac{\|T\|^2}{(\det T)^{2/n}}. \tag{3.22}$$

Notice that even if $T$ is not unique, the normalized size is defined up to universal multiplicative constants. In case $u \in C^2$ since

$$u(y) = u(x) + \nabla u(x) \cdot (y - x) + \frac{1}{2} D^2 u(x_0)(y - x) \cdot (y - x) + o(|y - x|^2),$$

we see that

$$\frac{1}{\sqrt{t}} S(x, t) \to \left\{ y : \quad \frac{1}{2} D^2 u(x_0)(y - x) \cdot (y - x) \le 1 \right\}.$$

Thus for $t$ small

$$\alpha(S(x, t)) \approx \|D^2 u(x)\|. \tag{3.23}$$

**Lemma 3.9.** *In the hypothesis of Theorem 3.8 we see that there exist positive universal constants $\bar{C}$ and $\bar{\beta}$ such that for all $x \in B_{3/4}$*

$$\operatorname{diam}(S(x, t)) \le \frac{\bar{C}}{(\alpha(S(x, t)))^{\bar{\beta}}} \qquad \forall t \le 1/\bar{C}. \tag{3.24}$$

*Proof.* Let $E$ be the John ellipsoid associate to $S(x, t)$ and let $\lambda_1 \le \cdots \le \lambda_n$ the length of its semi-axes. Since for $t$ universally small and $C$ universal by (2.21),

$$B(x, t/C) \subset S(x, t) \subset nE,$$

$n\lambda_1 \ge t/C$. Moreover for $C$ and $\beta$ universal, always by (2.21),

$$E \subset S(x, t) \subset B(x, Ct^\beta)$$

which implies $\lambda_n \le Ct^\beta \le \tilde{C}\lambda_1^\beta$. If $T$ is such that $T(E) = B_1$,

$$\alpha(S(x, t)) = \frac{\|T\|^2}{(\det T)^{2/n}} = \frac{(\lambda_1 \ldots \lambda_n)^{2/n}}{\lambda_1^2} \le \frac{\lambda_n^2}{\lambda_1^2} \le C\lambda_1^{2(\beta-1)}.$$

Thus, since $\beta < 1$,

$$\operatorname{diam}(S(x, t)) \le 2n\lambda_n \le C\lambda_1^\beta \le \frac{\bar{C}}{(\alpha(S(x, t)))^{\bar{\beta}}}$$

with $\bar{\beta} = \beta/(2 - 2\beta)$. $\qquad\square$

### 3.2.1. A direct proof of Theorem 3.8

**Lemma 3.10.** *Let $v$ be a normalized solution (see (2.12)), then there exists a universal constant $C_0$ such that*

$$\int_{\tilde{S}_{1/2}} \|D^2 v\| \leq C_0 \left| \{ C_0^{-1} \operatorname{Id} \leq D^2 v \leq C_0 \operatorname{Id} \} \cap \tilde{S}_{1/2\theta} \right|,$$

*where $\theta$ is the engulfing constant in Proposition 2.12 and, for all $\tau \in (0, 1)$,*

$$\tilde{S}_\tau = \{ x : v(x) \leq (1 - \tau) \min v \}$$

*Proof.* Exactly as in the proof of Lemma (3.5), we have

$$\int_{\tilde{S}_{1/2}} \|D^2 v\| \leq \int_{\tilde{S}_{1/2}} \Delta v = \int_{\partial \tilde{S}_{1/2}} v_\nu \leq C_1, \qquad (3.25)$$

where the last inequality follows from the interior Lipschitz estimate of $v$ in $\tilde{S}_{1/2}$. Recall that by Remark 2.7

$$\tilde{S}_{1/2} \geq c_1$$

for some $c_1 > 0$ universal. The last two inequalities show that the set

$$\left\{ \|D^2 v\| \leq 2C_1 c_1^{-1} \right\} \cap \tilde{S}_{1/2}$$

has at least measure $c_1/2$ in $S_{1/2\theta}$. Finally, the lower bound on $\det D^2 u$ implies that

$$C_0^{-1} \operatorname{Id} \leq D^2 v \leq C_0 \operatorname{Id} \qquad \text{inside } \{ \|D^2 v\| \leq 2C_1 c_1^{-1} \},$$

and the conclusion follows provided that we choose $C_0$ sufficiently large. $\square$

By rescaling we obtain:

**Lemma 3.11.** *Assume $S(x, 2t) \subset\subset \mathcal{U}$. If*

$$S(x, t) \text{ has normalized size } \overline{\alpha},$$

*then*

$$\int_{S(x_0,t)} \|D^2 u\| \leq C_0 \overline{\alpha} \left| \{ C_0^{-1} \overline{\alpha} \leq \|D^2 u\| \leq C_0 \overline{\alpha} \} \cap S(x, t/\theta) \right|.$$

*Proof.* We first notice that (see for instance the end the proof of Lemma 3.6)

$$\alpha(S(x, 2t)) \approx \alpha(S(x, t)) = \overline{\alpha}. \tag{3.26}$$

The lemma follows by applying Lemma 3.10 to the normalized solution $v$ build up from $u$. More precisely if $T$ normalizes $S(x, 2t)$, defining, as usual,

$$v(z) = (\det T)^{2/n}\left(u(T^{-1}z) - \nabla u(x) \cdot (T^{-1}z - x) - 2t\right) \quad z \in T(S(x, 2t)),$$

we see that $T(S(x, t)) = \tilde{S}_{1/2}$, $T(S(x, t/\theta)) = \tilde{S}_{1/2\theta}$, and

$$D^2u(x) = \frac{T^*D^2v(Tx)T}{(\det T)^{2/n}}.$$

Thus, by definition of normalized size (3.22) and (3.26),

$$|\det T| \int_{S(x_0, h/2)} \|D^2u\| \leq \overline{\alpha} \int_{S(x_0, 1/2)} \|D^2v\|$$

and

$$\left\{C_0^{-1}\,\mathrm{Id} \leq D^2v \leq C_0\,\mathrm{Id}\right\} \subset T\left(\left\{C_0^{-1}\overline{\alpha} \leq \|D^2u\| \leq C_0\overline{\alpha}\right\}\right)$$

which implies that

$$\left|\left\{C_0^{-1}I \leq D^2v \leq C_0\alpha\right\} \cap \tilde{S}_{1/2\theta}\right|$$

is bounded above by

$$|\det T|\,\left|\left\{C_0^{-1}\overline{\alpha} \leq \|D^2u\| \leq C_0\overline{\alpha}\right\} \cap S(x, t/2\theta)\right|.$$

The conclusion follows now by applying Lemma 3.10 to $v$. $\quad\square$

Next, for some large $M$, we denote by $D_k$ the closed sets

$$D_k := \left\{\,\|D^2u(x)\| \geq M^k\right\} \cap B_{R_k}, \tag{3.27}$$

where $R_0 = 3/4$ and

$$R_k = R_{k-1} - \bar{C}C_0^{-\bar{\beta}}M^{-k\bar{\beta}},$$

where $\bar{C}$ and $\bar{\beta}$ are as in Lemma 3.9 and $C_0$ is as in Lemma 3.11. As we show now, Lemma 3.11 combined with a covering argument gives a geometric decay for $\int_{D_k} \|D^2u\|$.

**Lemma 3.12.** *If $M = C_2$, with $C_2$ a large universal constant, then*

$$\int_{D_{k+1}} \|D^2 u\| \, dx \leq (1 - \tau) \int_{D_k} \|D^2 u\| \, dx,$$

*for some small universal constant $\tau > 0$ and*

$$B_{1/2} \subset B_{R_k} \qquad \forall k \geq 0$$

*Proof.* Let $\rho$, universal, be such that $S(x, t) \Subset \mathcal{U}$ for all $x \in B_{3/4}$, for $t \leq \rho$ (cp. Proposition 2.12). Let $M \gg C_0$ (to be fixed later), and for each $x \in D_{k+1}$ consider a section

$$S(x, t) \text{ of normalized size } \alpha = C_0 M^k,$$

which is compactly included in $\mathcal{U}$. This is possible if $M$ is universally large, since for $t \to 0$ the normalized size of $S(x, t)$ is comparable to $\|D^2 u(x)\|$ (recall (3.23)) which is greater than $M^{k+1} > \alpha$, whereas if $t = \rho$ the normalized size is bounded above by a universal constant and therefore by $\alpha$.

Now we choose a Vitali cover for $D_{k+1}$ with sections $S(x_i, t_i)$, $i = 1, \ldots, m$. Notice that by Lemma 3.9

$$\text{diam}(S(x_i, t_i)) \leq \bar{C} C_0^{-\bar{\beta}} M^{-(k+1)\bar{\beta}},$$

thus

$$\bigcup_i S(x_i, t_i) \subset B_{R_k}. \tag{3.28}$$

By Lemma 3.11, for each $i$,

$$\int_{S(x_i, t_i)} \|D^2 u\| \leq C_0^2 M^k \left| \left\{ M^k \leq \|D^2 u\| \leq C_0^2 M^k \right\} \cap S(x_i, t_i/\theta) \right|.$$

Adding these inequalities and using

$$D_{k+1} \subset \bigcup S(x_i, t_i) \cap B_{R_k}, \qquad S(x_i, t_i/\theta) \text{ disjoint,}$$

and (3.28), we obtain

$$\int_{D_{k+1}} \|D^2 u\| \leq C_0^2 M^k \left| \left\{ M^k \leq \|D^2 u\| \leq C_0^2 M^k \right\} \cap B_{R_k} \right|$$

$$\leq C \int_{D_k \setminus D_{k+1}} \|D^2 u\| dx$$

provided $M \geq C_0^2$. Adding $C \int_{D_{k+1}} \|D^2 u\|$ to both sides of the above inequality, the first claim follows with $\tau = 1/(1 + C)$. To obtain the second one just choose $M$ universally large such that

$$\bar{C} \bar{C}_0^{\bar{\beta}} \sum_{k \geq 1} M^{-\bar{\beta} k} \leq 1/4. \tag{3.29}$$

$\square$

By the above result, the proof of (3.8) is immediate: indeed, by Lemma 3.12 we easily deduce that there exist $C, \varepsilon > 0$ universal such that

$$M^k |\{x \in B_{1/2} : \|D^2 u(x)\| \geq M^k\}| \leq \int_{\{x \in B_{1/2}: \|D^2 u(x)\| \geq M^k\}} \|D^2 u\|$$

$$\leq \int_{D_k} \|D^2 u\| \leq C M^{-2k\varepsilon}.$$

Since, by Fubini Theorem,

$$\int_{B_{1/2}} \|D^2 u\|^{1+\varepsilon} \approx \sum_k M^{(1+\varepsilon)k} |\{x \in B_{1/2} : \|D^2 u(x)\| \geq M^k\}|,$$

we obtain the proof of Theorem 3.8.

### 3.2.2. A proof by iteration of the $L \log L$ estimate

We now briefly sketch how Theorem 3.8 could also be easily deduced by applying the $L \log L$ estimate of Theorem 3.1 inside every section, and then doing a covering argument.

First for a normalized solution $v$, and $K > 0$ we introduce the notation

$$F_K := \{\|D^2 v\| \geq K\} \cap \tilde{S}_{1/2},$$

see Lemma 3.10.

**Lemma 3.13.** *Suppose $v$ satisfies the assumptions of Lemma 3.10. Then there exist universal constants $C_0$ and $C_1$ such that, for all $K \geq 2$,*

$$|F_K| \leq \frac{C_1}{K \log(K)} \left| \{C_0^{-1} \operatorname{Id} \leq D^2 u \leq C_0 \operatorname{Id}\} \cap \tilde{S}_{1/2\theta} \right|.$$

Indeed, from the proof of Lemma 3.10 the measure of the set appearing on the right hand side is bounded below by a small universal constant $c_1/2$, while by Theorem 3.4 $|F_K| \leq C/K \log(K)$ for all $K \geq 2$, hence

$$|F_K| \leq \frac{2C}{c_1 K \log(K)} \left| \{C_0^{-1} \operatorname{Id} \leq D^2 v \leq C_0 \operatorname{Id}\} \cap \tilde{S}_{1/2\theta} \right|.$$

Exactly as in the proof of Lemma 3.11, by rescaling we obtain:

**Lemma 3.14.** *Suppose u satisfies the assumptions of Lemma* 3.11. *Then,*

$$|\{\|D^2 u\| \geq \overline{\alpha} K\} \cap S(x, t)| \leq \frac{C_1}{K \log(K)} \left|\{C_0^{-1}\overline{\alpha} \leq \|D^2 u\|\} \cap S(x, t/\theta)\right|,$$

*for all* $K \geq 2$.

Finally, as proved in the next Lemma, a covering argument shows that the measure of the sets $D_k$ defined in (3.27) decays as $M^{-(1+2\varepsilon)k}$, which shows the desired integrability.

**Lemma 3.15.** *There exist universal constants M large and $\varepsilon > 0$ small such that*

$$|D_{k+1}| \leq M^{-1-2\varepsilon}|D_k|.$$

*Proof.* As in the proof of Lemma 3.12, we use a Vitali covering of the set $D_{k+1}$ with sections $S(x, t)$ of normalized size $\alpha = C_0 M^k$, i.e.

$$D_{k+1} \subset \bigcup S(x_i, t_i), \qquad S(x_i, t_i/\theta) \text{ disjoint sets.}$$

Apply Lemma 3.14 above for

$$K := C_0^{-1} M,$$

hence $\alpha K = M^{k+1}$, and find that for each $i$

$$|D_{k+1} \cap S(x_i, t_i)| \leq \frac{2C_0}{M \log(M)} |D_k \cap S(x_i, t_i/\theta)|,$$

provided that $M \gg C_0$. Summing over $i$ and choosing $M \geq e^{4C_0}$ we get

$$|D_{k+1}| \leq \frac{2C_0}{M \log(M)} |D_k| \leq \frac{1}{2M} |D_k|,$$

and the lemma is proved by choosing $\varepsilon = \log(2)/\log(M)$. □

Finally if we choose $M$ so large such that also (3.29) is satisfied, the above Lemma implies

$$|\{x \in B_{1/2} : \|D^2 u(x)\| \geq M^k\}| \leq M^{-k(1+2\varepsilon)},$$

and we conclude as in the previous section.

## 3.3. A simple proof of Caffarelli $W^{2,p}$ estimates

Using the technique of the previous Section we give a simple proof of celebrated Caffarelli $W^{2,p}$ estimates (see Theorem 2.20). More precisely we will prove the following

**Theorem 3.16.** *For all $p > 1$ there exist a $\delta_p$ and a constant $C_p$ such that: if $u : \overline{\mathcal{U}} \to \mathbb{R}$ is a $C^2$ solution of (2.1) with $\|f - 1\| \leq \delta_p$ and $B_1 \subset \mathcal{U} \subset B_n$, then*

$$\int_{B_{1/2}} \|D^2 u\|^p \leq C_p. \tag{3.30}$$

The proof, which is briefly sketched in [53], is, as the original one, again based on a decay estimate of the type

$$|\{x \in B_{1/2} : \|D^2 u(x)\| \geq M^k\}| \lesssim M^{-k(1+p)}. \tag{3.31}$$

Again we start with a "normalized" lemma:

**Lemma 3.17.** *Let $v$ be a normalized solution (see (2.12)), then there exists a universal constant $C_0$ such that for every $\eta$ there exists a $\delta = \delta(\eta)$ such that if $\|f - 1\| \leq \delta$*

$$\left|\{D^2 v \geq C_0 \, \mathrm{Id}\} \cap \tilde{S}_{1/2}\right| \leq C_0 \eta \, \left|\{C_0^{-1} \, \mathrm{Id} \leq D^2 v \leq C_0 \, \mathrm{Id}\} \cap \tilde{S}_{1/2\theta}\right|,$$

*where $\theta$ is the engulfing constant in Proposition 2.12 and $\tilde{S}_\tau$ is as in Lemma 3.10.*

The above Lemma is well known and it was a key step in the proof of [19], see also [67], and the dependence of $\delta$ from $\eta$ can be also quantified in a power like one ($\delta \lesssim \eta^\gamma$). Here we give a proof based on the results of Chapter 4 (which are independent on the above lemma).

*Proof.* Again by the proof Lemma 3.10, the measure of the set appearing on the right hand side is bounded below by a small universal constant $c_1/2$. It will be hence enough to show that for some universal constant $C_0$

$$\left|\{D^2 v \geq C_0 \, \mathrm{Id}\} \cap \tilde{S}_{1/2}\right| \to 0$$

as $\delta \to 0$ (uniformly in $v$). Choose a sequence $\delta_k \to 0$ and let $v_k$ be a sequence of normalized solutions defined on normalized sets $Z_k$. By Lemma 2.10, $Z_k$ converge in the Hausdorff distance to a normalized set $Z_\infty$ and $v_k$ uniformly converge in $Z_\infty$ to $v_\infty$, a normalized solution of

$$\begin{cases} \det D^2 v_\infty = 1 & \text{in } Z_\infty \\ v_\infty = 0 & \text{on } \partial Z_\infty. \end{cases}$$

Thanks to Theorem 4.1 $v_k \to v_\infty$ in $W^{2,1}_{loc}(Z_\infty)$. Since (with obvious notations) $\tilde{S}^k_{1/2} \to \tilde{S}^\infty_{1/2}$ they are definitely well contained in $\tilde{S}^\infty_{3/4}$, hence

$$\int_{\tilde{S}^k_{1/2}} \|D^2 v_k - D^2 v_\infty\| \leq \int_{\tilde{S}^\infty_{3/4}} \|D^2 v_k - D^2 v_\infty\| \to 0.$$

By Theorem 2.16, we know that the $C^2$ norm of $v_\infty$ on $S^\infty_{3/4}$ is bounded from above by a universal constant $C_1$, thus

$$\left| \{D^2 v_k \geq 4C_1 \, \mathrm{Id}\} \cap \tilde{S}^k_{1/2} \right|$$

$$\leq \left| \{\|D^2 v_\infty\| \geq 2C_1\} \cap \tilde{S}^\infty_{3/4} \right| + \left| \{\|D^2 v_k - D^2 v_\infty\| \geq 2C_1\} \cap \tilde{S}^k_{1/2} \right|$$

$$= \left| \{\|D^2 v_k - D^2 v_\infty\| \geq 2C_1\} \cap \tilde{S}^k_{1/2} \right|$$

$$\leq \frac{1}{2C_1} \int_{\tilde{S}^k_{1/2}} \|D^2 v_k - D^2 v_\infty\| \to 0,$$

proving the claim with $C_0 \geq 4C_1$. □

Exactly as in the proof of Lemma 3.11 we obtain

**Lemma 3.18.** *Assume $S(x, 2t) \Subset \mathcal{U}$. If*

$$S(x, t) \text{ has normalized size } \overline{\alpha},$$

*then for all $\eta$ there exists a $\delta = \delta(\eta)$ such that if $\|f - 1\|_\infty \leq \delta$ then*

$$|\{\|D^2 u\| \geq \overline{\alpha} C_0\} \cap S(x, t)| \leq C_0 \eta \left| \{C_0^{-1} \overline{\alpha} \leq \|D^2 u\|\} \cap S(x, t/\theta) \right|,$$

*where $C_0$ is as in Lemma 3.17.*

Defining $D_k$ as in (3.27) (with $M$ so big to satisfy (3.29)) and arguing as in the proof of Lemma 3.15 we see that

$$|D_{k+1}| \leq C_0 \eta |D_k|,$$

from which it follows

$$|\{x \in B_{1/2} : \|D^2 u(x)\| \geq M^k\}| \leq (C_0 \eta)^{-k}.$$

Since $M$ is universally fixed, we can choose $\eta$ (and hence $\delta$) such that $C_0 \eta = M^{-(p+1)}$, obtaining the inequality (3.31).

# Chapter 4
# Second order stability for the Monge-Ampère equation and applications

A question which naturally arises in view of the previous results (and which has been suggested to us by Luigi Ambrosio) is the following: choose a sequence of functions $f_k$ with $\lambda \leq f_k \leq 1/\lambda$ which converges to $f$ strongly in $L^1_{\text{loc}}(\Omega)$, and denote by $u_k$ and $u$ the solutions of (2.1) corresponding to $f_k$ and $f$ respectively. By the convexity of $u_k$ and $u$, and the uniqueness of solutions to (2.1), it is immediate to deduce that $u_k \rightarrow u$ uniformly, and $\nabla u_k \rightarrow \nabla u$ in $L^p_{\text{loc}}(\Omega)$ for any $p < \infty$. What can be said about the strong convergence of $D^2 u_k$? Due to the highly nonlinear character of the Monge-Ampère equation, this question is nontrivial.

In this chapter we report the results of [41] where, in collaboration with Alessio Figalli, we addressed this problem. Our main results are the following

**Theorem 4.1.** *Let $\Omega_k \subset \mathbb{R}^n$ be convex domains, and let $u_k : \Omega_k \rightarrow \mathbb{R}$ be convex Aleksandrov solutions of*

$$\begin{cases} \det D^2 u_k = f_k & in\ \Omega_k \\ u_k = 0 & on\ \partial\Omega_k \end{cases} \tag{4.1}$$

*with $0 < \lambda \leq f_k \leq 1/\lambda$. Assume that $\Omega_k$ converge to some convex domain $\Omega$ in the Hausdorff distance, and $f_k \chi_{\Omega_k}$ converge to $f$ in $L^1_{\text{loc}}(\Omega)$. Then, if $u$ denotes the unique Aleksandrov solution of*

$$\begin{cases} \det D^2 u = f & in\ \Omega \\ u = 0 & on\ \partial\Omega, \end{cases}$$

*for any $\Omega' \Subset \Omega$ we have*

$$\|u_k - u\|_{W^{2,1}(\Omega')} \rightarrow 0 \qquad as\ k \rightarrow \infty. \tag{4.2}$$

Obviously, since the functions $u_k$ are uniformly bounded in $W^{2,\gamma_0}(\Omega')$, this gives strong convergence in $W^{2,\gamma'}(\Omega')$ for any $\gamma' < \gamma_0$.

The consequences for what concerns optimal transportation are summarized in the following theorem:

**Theorem 4.2.** *Let* $\Omega_1, \Omega_2 \subset \mathbb{R}^n$ *be two bounded domains with* $\Omega_2$ *convex, and let* $f_k, g_k$ *be a family of probability densities such that* $0 < \lambda \leq f_k, g_k \leq 1/\lambda$ *inside* $\Omega_1$ *and* $\Omega_2$ *respectively. Assume that* $f_k \to f$ *in* $L^1(\Omega_1)$ *and* $g_k \to g$ *in* $L^1(\Omega_2)$, *and let* $T_k : \Omega_1 \to \Omega_2$ *(resp.* $T : \Omega_1 \to \Omega_2$) *be the (unique) optimal transport map for the quadratic cost sending* $f_k$ *onto* $g_k$ *(resp.* $f$ *onto* $g$). *Then* $T_k \to T$ *in* $W_{\text{loc}}^{1,\gamma'}(\Omega_1)$ *for some* $\gamma' > 1$.

If $v : \overline{\Omega} \to \mathbb{R}$ is a continuous function, we define its *convex envelope inside* $\Omega$ as

$$\Gamma_v(x) := \sup\{\ell(x) : \ell \leq v \text{ in } \Omega, \ \ell \text{ affine}\}. \tag{4.3}$$

In case $\Omega$ is a convex domain and $v \in C^2(\Omega)$, it is easily seen that

$$D^2 v(x) \geq 0 \qquad \text{for every } x \in \{v = \Gamma_v\} \cap \Omega \tag{4.4}$$

in the sense of symmetric matrices. Moreover the following inequality between measures holds in $\Omega$ ($\mu_{\Gamma_v}$ is the Monge-Ampère measure associated to $\Gamma_v$, see Section 1.2):

$$\mu_{\Gamma_v} \leq \det D^2 v \mathbf{1}_{\{v = \Gamma_v\}} \, dx. \tag{4.5}$$

To see this, let us first recall that it is well known that if $x_0 \in \Omega \setminus \{\Gamma_v = v\}$ and $p \in \partial \Gamma_v(x_0)$ then the convex set

$$\{x \in \Omega : \Gamma_v(x) = p \cdot (x - x_0) + \Gamma_v(x_0)\}$$

is nonempty and contains more than one point (see for instance [42]). In particular

$$\partial \Gamma_v\big(\Omega \setminus \{\Gamma_v = v\}\big)$$
$$\subset \{p \in \mathbb{R}^n : \text{ there exist } x, y \in \Omega, x \neq y \text{ and } p \in \partial \Gamma_v(x) \cap \partial \Gamma_v(y)\}.$$

and by Lemma 1.17 this last set has measure zero. Hence

$$\big|\partial \Gamma_v\big(\Omega \setminus \{\Gamma_v = v\}\big)\big| = 0. \tag{4.6}$$

Moreover, since $v \in C^1(\Omega)$, for any $x \in \{\Gamma_v = v\} \cap \Omega$ it holds $\partial \Gamma_v(x) = \{\nabla v(x)\}$. Thus, using (4.6) and (4.4), for any open set $A \Subset \Omega$ we have

$$\mu_{\Gamma_v}(A) = \big|\partial \Gamma_v\big(A \cap \{\Gamma_v = v\}\big)\big| = \big|\nabla v\big(A \cap \{\Gamma_v = v\}\big)\big|$$
$$\leq \int_{A \cap \{\Gamma_v = v\}} |\det D^2 v| = \int_{A \cap \{\Gamma_v = v\}} \det D^2 v.$$

(The inequality above follows from the Area Formula (1.15) applied to the $C^1$ map $\nabla v$.) This proves (4.5).

We also recall (see Appendix A) that a continuous function $v$ is said to be *twice differentiable* at $x$ if there exists a (unique) vector $\nabla v(x)$ and a (unique) symmetric matrix $\nabla^2 v(x)$ such that

$$v(y) = v(x) + \nabla v(x) \cdot (y - x) + \frac{1}{2}\nabla^2 v(x)(y - x) \cdot (y - x) + o(|y - x|^2).$$

In case $v$ is twice differentiable at some point $x_0 \in \{v = \Gamma_v\}$, then it is immediate to check that

$$\nabla^2 v(x_0) \geq 0. \tag{4.7}$$

By Aleksandrov Theorem, any convex function is twice differentiable almost everywhere. In particular (4.7) holds almost everywhere on $\{v = \Gamma_v\}$, whenever $v$ is the difference of two convex functions.

Finally we recall that, in case $v \in W_{\text{loc}}^{2,1}$, then the pointwise Hessian of $v$ coincides almost everywhere with its distributional Hessian (cp. Appendix A). Since in the sequel we are going to deal with $W_{\text{loc}}^{2,1}$ convex functions, we will use $D^2 u$ to denote both the pointwise and the distributional Hessian.

## 4.1. Proof of Theorem 4.1

We are going to use the following result:

**Lemma 4.3.** *Let* $\Omega \subset \mathbb{R}^n$ *be a strictly convex bounded domain, and let* $u, v : \overline{\Omega} \to \mathbb{R}$ *be continuous strictly convex functions such that* $\mu_u = f\mathscr{L}^n$ *and* $\mu_v = g\mathscr{L}^n$, *with* $f, g \in L_{\text{loc}}^1(\Omega)$. *Then*

$$\mu_{\Gamma_{u-v}} \leq \left(f^{1/n} - g^{1/n}\right)^n \mathbf{1}_{\{u-v=\Gamma_{u-v}\}} \, dx. \tag{4.8}$$

*Proof.* In case $u$, $v$ are of class $C^2(\Omega)$, by (4.4) we have

$$0 \leq D^2 u(x) - D^2 v(x) \qquad \text{for every } x \in \{u - v = \Gamma_{u-v}\},$$

so using the monotonicity and the concavity of the function $\det^{1/n}$ on the cone of non-negative symmetric matrices we get

$$0 \leq \det(D^2 u - D^2 v)$$
$$\leq \left(\left(\det D^2 u\right)^{1/n} - \left(\det D^2 v\right)^{1/n}\right)^n \qquad \text{on } \{u - v = \Gamma_{u-v}\},$$

which combined with (4.5) gives the desired result.

Now, for the general case, we consider two sequences of smooth functions $f_k$ and $g_k$ converging respectively to $f$ and $g$ in $L^1(\Omega)$, and we solve (see [67, Theorem 1.6.2])

$$\begin{cases} \det D^2 u_k = f_k & \text{in } \Omega \\ u_k = \bar{u}_k & \text{on } \partial\Omega, \end{cases} \qquad \begin{cases} \det D^2 v_k = g_k & \text{in } \Omega \\ v_k = \bar{v}_k & \text{on } \partial\Omega, \end{cases}$$

Where $\bar{u}_k$ (reps. $\bar{v}_k$) are smooth approximations of of $u$ (reps. $v$) on $\partial\Omega$. In this way $u_k$ (reps. $v_k$) are smooth on $\Omega$ (see Section 2.3) and, by [67, Lemma1.6.1], continuous on $\overline{\Omega}$ with a modulus of continuity which depends only on $u$ (resp. $v$). Hence they converge uniformly on $\overline{\Omega}$ to $u$ (reps $v$). Thus, also $\Gamma_{u_k - v_k}$ converges uniformly in $\overline{\Omega}$ to $\Gamma_{u-v}$ (this easily follows by the definition of convex envelope). Moreover, it follows easily from the definition of contact set that

$$\limsup_{k \to \infty} \mathbf{1}_{\{u_k - v_k = \Gamma_{u_k - v_k}\}} \leq \mathbf{1}_{\{u - v = \Gamma_{u-v}\}}. \tag{4.9}$$

We now observe that the previous step applied to $u_k$ and $v_k$ gives

$$\mu_{\Gamma_{u_k - v_k}} \leq \left( \left( \det D^2 u_k \right)^{1/n} - \left( \det D^2 v_k \right)^{1/n} \right)^n \mathbf{1}_{\{u_k - v_k = \Gamma_{u_k - v_k}\}} \, dx,$$

Thus, letting $k \to \infty$ and taking in account Proposition 1.25 and (4.9), we obtain (4.8). $\qquad\square$

*Proof of Theorem* 4.1. The $L^1_{\text{loc}}$ convergence of $u_k$ (resp. $\nabla u_k$) to $u$ (resp. $\nabla u$) is easy and standard, see Lemma 2.10, so we focus on the convergence of the second derivatives.

Without loss of generality we can assume that $\Omega'$ is strictly convex, and that $\Omega' \Subset \Omega_k$ (since $\Omega_k \to \Omega$ in the Hausdorff distance, this is always true for $k$ sufficiently large). Fix $\varepsilon \in (0, 1)$, let $\Gamma_{u - (1-\varepsilon)u_k}$ be the convex envelope of $u - (1 - \varepsilon)u_k$ inside $\Omega'$ (see (4.3)), and define

$$A_k^\varepsilon := \{ x \in \Omega' : u(x) - (1 - \varepsilon)u_k(x) = \Gamma_{u - (1-\varepsilon)u_k}(x) \}.$$

Since $u_k \to u$ uniformly in $\overline{\Omega}'$, $\Gamma_{u-(1-\varepsilon)u_k}$ converges uniformly to $\Gamma_{\varepsilon u} = \varepsilon u$ (as $u$ is convex) inside $\Omega'$. Hence, by applying Proposition 1.25 and (4.8) to $u$ and $(1 - \varepsilon)u_k$ inside $\Omega'$, we get that

$$\varepsilon^n \int_{\Omega'} f = \mu_{\Gamma_{\varepsilon u}}(\Omega')$$
$$\leq \liminf_{k \to \infty} \mu_{\Gamma_{u-(1-\varepsilon)u_k}}(\Omega')$$
$$\leq \liminf_{k \to \infty} \int_{\Omega' \cap A_k^\varepsilon} \left( f^{1/n} - (1 - \varepsilon)f_k^{1/n} \right)^n.$$

We now observe that, since $f_k$ converges to $f$ in $L^1_{\text{loc}}(\Omega)$, we have

$$\left| \int_{\Omega' \cap A^\varepsilon_k} \left( f^{1/n} - (1 - \varepsilon) f_k^{1/n} \right)^n - \int_{\Omega' \cap A^\varepsilon_k} \varepsilon^n f \right|$$

$$\leq \int_{\Omega'} \left| \left( f^{1/n} - (1 - \varepsilon) f_k^{1/n} \right)^n - \varepsilon^n f \right| \to 0$$

as $k \to \infty$. Hence, combining the two estimates above, we immediately get

$$\int_{\Omega'} f \leq \liminf_{k \to \infty} \int_{\Omega' \cap A^\varepsilon_k} f,$$

or equivalently

$$\limsup_{k \to \infty} \int_{\Omega' \setminus A^\varepsilon_k} f = 0.$$

Since $f \geq \lambda$ inside $\Omega$ (as a consequence of the fact that $f_k \geq \lambda$ inside $\Omega_k$), this gives

$$\lim_{k \to \infty} |\Omega' \setminus A^\varepsilon_k| = 0 \qquad \forall \varepsilon \in (0, 1). \tag{4.10}$$

We now recall that, thanks to Theorems 2.11 and 3.1, $u_k$ are strictly convex and belongs to $W^{2,1}(\Omega')$. Hence we can apply (4.7) to deduce that

$$D^2 u - (1 - \varepsilon) D^2 u_k \geq 0 \qquad \text{a.e. on } A^\varepsilon_k.$$

In particular, by (4.10),

$$|\Omega' \setminus \{ D^2 u \geq (1 - \varepsilon) D^2 u_k \}| \to 0 \qquad \text{as } k \to \infty.$$

By a similar argument (exchanging the roles of $u$ and $u_k$)

$$|\Omega' \setminus \{ (1 - \varepsilon) D^2 u \leq D^2 u_k \}| \to 0 \qquad \text{as } k \to \infty.$$

Hence, if we call $B^\varepsilon_k := \left\{ x \in \Omega' : (1 - \varepsilon) D^2 u_k \leq D^2 u \leq \frac{1}{1-\varepsilon} D^2 u_k \right\}$, it holds

$$\lim_{k \to \infty} |\Omega' \setminus B^\varepsilon_k| = 0 \qquad \forall \varepsilon \in (0, 1).$$

Moveover, by (3.2) applied to both $u_k$ and $u$, we have

$$\int_{\Omega'} \| D^2 u - D^2 u_k \| = \int_{\Omega' \cap B^\varepsilon_k} \| D^2 u - D^2 u_k \| + \int_{\Omega' \setminus B^\varepsilon_k} \| D^2 u - D^2 u_k \|$$

$$\leq \frac{\varepsilon}{1 - \varepsilon} \int_{\Omega'} \| D^2 u \| + \| D^2 u - D^2 u_k \|_{L^{\gamma_0}(\Omega')} |\Omega' \setminus B^\varepsilon_k|^{1 - 1/\gamma_0}$$

$$\leq C \left( \frac{\varepsilon}{1 - \varepsilon} + |\Omega' \setminus B^\varepsilon_k|^{1 - 1/\gamma_0} \right).$$

Hence, letting first $k \to \infty$ and then sending $\varepsilon \to 0$, we obtain the desired result. $\qquad\square$

## 4.2. Proof of Theorem 4.2

In order to prove Theorem 4.2, we will need the following lemma (note that for the next result we do not need to assume convexity of the target domain):

**Lemma 4.4.** *Let $\Omega_1, \Omega_2 \subset \mathbb{R}^n$ be two bounded domains, and let $f_k, g_k$ be probability densities such that $0 < \lambda \le f_k, g_k \le 1/\lambda$ inside $\Omega_1$ and $\Omega_2$ respectively. Assume that $f_k \to f$ in $L^1(\Omega_1)$ and $g_k \to g$ in $L^1(\Omega_2)$, and let $T_k : \Omega_1 \to \Omega_2$ (resp. $T : \Omega_1 \to \Omega_2$) be the (unique) optimal transport map for the quadratic cost sending $f_k$ onto $g_k$ (resp. $f$ onto $g$). Then*

$$\frac{f_k}{g_k \circ T_k} \to \frac{f}{g \circ T} \qquad in\ L^1(\Omega_1).$$

*Proof.* By stability of optimal transport maps (see for instance Theorem 1.14) and the fact that $f_k \ge \lambda$ (and so $f \ge \lambda$), we know that $T_k \to T$ in measure (with respect to Lebesgue) inside $\Omega$.

We claim that $\varphi \circ T_k \to \varphi \circ T$ in $L^1(\Omega_1)$ for all $\varphi \in L^\infty(\Omega_2)$. Indeed this is obvious if $\varphi$ is uniformly continous (by the convergence in measure of $T_k$ to $T$). In the general case we choose $\varphi_\eta \in C(\overline{\Omega}_2)$ such that $\|\varphi - \varphi_\eta\|_{L^1(\Omega_2)} \le \eta$ and we observe that (recall that $f_k, f \ge \lambda$, $g_k, g \le 1/\lambda$, and $T_\# f_k = g_k$, $T_\# f = g$)

$$\int_{\Omega_1} |\varphi \circ T_k - \varphi \circ T| \le \int_{\Omega_1} |\varphi_\eta \circ T_k - \varphi_\eta \circ T|$$
$$+ \int_{\Omega_1} |\varphi_\eta \circ T_k - \varphi \circ T_k| \frac{f_k}{\lambda}$$
$$+ \int_{\Omega_1} |\varphi_\eta \circ T - \varphi \circ T| \frac{f}{\lambda}$$
$$= \int_{\Omega_1} |\varphi_\eta \circ T_k - \varphi_\eta \circ T| + \int_{\Omega_2} |\varphi_\eta - \varphi| \frac{g_k}{\lambda}$$
$$+ \int_{\Omega_2} |\varphi_\eta - \varphi| \frac{g}{\lambda}$$
$$\le \int_{\Omega_1} |\varphi_\eta \circ T_k - \varphi_\eta \circ T| + \frac{2\eta}{\lambda^2}.$$

Thus

$$\limsup_{k\to\infty} \int_{\Omega_1} |\varphi \circ T_k - \varphi \circ T| \le \frac{2\eta}{\lambda^2},$$

and the claim follows by the arbitrariness of $\eta$.

Since

$$\int_{\Omega_1} |g_k \circ T_k - g \circ T| \leq \int_{\Omega_1} |g_k \circ T_k - g \circ T_k| \frac{f_k}{\lambda} + \int_{\Omega_1} |g \circ T_k - g \circ T|$$

$$= \int_{\Omega_2} |g_k - g| \frac{g_k}{\lambda} + \int_{\Omega_1} |g \circ T_k - g \circ T|$$

$$\leq \frac{1}{\lambda^2} \|g_k - g\|_{L^1(\Omega_2)} + \int_{\Omega_1} |g \circ T_k - g \circ T|,$$

from the claim above with $\varphi = g$ we immediately deduce that also $g_k \circ T_k \to g \circ T$ in $L^1(\Omega_1)$.

Thanks to $\lambda \leq g_k \circ T_k \leq 1/\lambda$ and $g_k \circ T_k \to g \circ T$ in $L^1(\Omega)$, it is immediate to see that

$$\frac{1}{g_k \circ T_k} \to \frac{1}{g \circ T} \qquad \text{in } L^p(\Omega_1) \text{ for every } p \in (1, +\infty).$$

Since also $f_k \to f$ in $L^p(\Omega_1)$ for every $p$, it follows that

$$\frac{f_k}{g_k \circ T_k} \to \frac{f}{g \circ T} \qquad \text{in } L^1(\Omega_1),$$

which is the desired result.                                                    □

*Proof of Theorem* 4.2. Since $T_k$ are uniformly bounded in $W^{1,\gamma_0}(\Omega_1')$ for some $\gamma_0 > 1$ and any $\Omega_1' \Subset \Omega$, it suffices to prove that $T_k \to T$ in $W^{1,1}_{\text{loc}}(\Omega_1)$.

Fix $x_0 \in \Omega_1$ and $r > 0$ such that $B_r(x_0) \Subset \Omega_1$. By compactness, it suffices to show that there is an open neighborhood $\mathcal{U}_{x_0}$ of $x_0$ such that $\mathcal{U}_{x_0} \subset B_r(x_0)$ and

$$\int_{\mathcal{U}_{x_0}} |T_k - T| + |\nabla T_k - \nabla T| \to 0.$$

By Theorems 1.8 and 2.2 the maps $T_k$ (resp. $T$) can be written as $\nabla u_k$ (resp. $\nabla u$) for some strictly convex function $u_k : B_r(x_0) \to \mathbb{R}$ (resp. $u : B_r(x_0) \to \mathbb{R}$). Moreover, possibly subtracting an additive constant (which will change neither $T_k$ nor $T$), one may assume that $u_k(x_0) = u(x_0)$.

Since the maps $T_k = \nabla u_k$ are bounded (as they take values in the bounded set $\Omega_2$), by Theorem 1.14 we get that $\nabla u_k \to \nabla u$ in $L^1_{\text{loc}}(B_r(x_0))$. (Actually, if one uses Theorem 2.1, $\nabla u_k$ are locally uniformly Hölder maps, so they converge locally uniformly to $\nabla u$.) Hence, to conclude the proof we only need to prove the convergence of $D^2 u_k$ to $D^2 u$ in a neighborhood of $x_0$.

To this aim, we observe that, by strict convexity of $u$ (see Theorem 2.2), we can find a linear function $\ell(z) = a \cdot z + b$ such that the open convex set $Z := \{z : u(z) < u(x_0) + \ell(z)\}$ is non-empty and compactly supported inside $B_{r/2}(x_0)$. Hence, by the uniform convergence of $u_k$ to $u$ (which follows from the $L^1_{loc}$ convergence of the gradients, the convexity of $u_k$ and $u$, and the fact that $u_k(x_0) = u(x_0)$), and the fact that $\nabla u$ is transversal to $\ell$ on $\partial Z$, we get that $Z_k := \{z : u_k(z) < u_k(x_0) + \ell(z)\}$ are non-empty convex sets which converge in the Hausdorff distance to $Z$.

Moreover, by Proposition 1.23 the maps $v_k := u_k - \ell$ solve in the Aleksandrov sense

$$\begin{cases} \det D^2 v_k = \frac{f_k}{g_k \circ T_k} & \text{in } Z_k \\ v_k = 0 & \text{on } \partial Z_k. \end{cases}$$

Therefore, thanks to Lemma 4.4 we can apply Theorem 4.1 to deduce that $D^2 u_k \to D^2 u$ in any relatively compact subset of $Z$, which concludes the proof. $\qquad \square$

# Chapter 5
# The semigeostrophic equations

In this Chapter we use the Sobolev regularity of optimal transport maps proved in Chapter 3 to show the existence of distributional solutions of the semigeostrophic equations, a simple model used in meteorology to describe large scale atmospheric flows.

The Chapter[1] is structured as follows: in Section 5.1 we introduce the model and show how, thanks to a change of variable due to Hoskins, there is a natural "dual equation" associated to it. This will also reveal the link between the semigeostrophic equations and optimal transoprtation. In Section 5.2 we discuss the case of periodic 2-dimensional solutions, finally in Section 5.3 we discuss the case of 3-dimensional flows.

## 5.1. The semigeostrophic equations in physical and dual variables

As explained for instance in [13, Section 2.2] and [77, Section 1.1] (see also [34] and [81] for a more complete exposition), the semigeostrophic equations can be derived from the 3-d Euler equations, with Boussinesq and hydrostatic approximations, subject to a strong Coriolis force.

More precisely, with the appropriate choice of units, the semigeo-strophic equations can be written, on a domain $\Omega$, as

$$\begin{cases} \partial_t u_t^g + \left( u_t \cdot \nabla \right) u_t^g + \nabla p_t = -J u_t + m_t e_3 & \text{in } \Omega \times (0, \infty) \\ u_t^g = J \nabla p_t \\ \partial_t m_t + \left( u_t \cdot \nabla \right) m_t = 0 & \text{in } \Omega \times (0, \infty) \\ \nabla \cdot u_t = 0 & \text{in } \Omega \times [0, \infty) \\ u_t \cdot \nu_\Omega = 0 & \text{in } \partial\Omega \times [0, \infty) \\ p_0 = p^0 & \text{in } \Omega. \end{cases} \tag{5.1}$$

---

[1] based on [5, 6] in collaboration with Luigi Ambrosio, Maria Colombo and Alessio Figalli.

Here $p^0$ is the initial condition for $p$, $\nu_\Omega$ is the unit outward normal to $\partial\Omega$, $e_3 = (0, 0, 1)^T$ is the third vector of the canonical basis in $\mathbb{R}^3$, $J$ is the matrix given by

$$J := \begin{pmatrix} 0 & -1 & 0 \\ 1 & 0 & 0 \\ 0 & 0 & 0 \end{pmatrix},$$

and the functions $u_t$, $p_t$, and $m_t$ represent respectively the *velocity*, the *pressure* and the *buoyancy* of the atmosphere, while $u_t^g$ is the so-called *semi-geostrophic* wind.[2] Clearly the pressure is defined up to a (time-dependent) additive constant.

For large scale atmospheric flows the Coriolis force dominates the advection term, hence the flow is mostly bi-dimensional, notice in fact that the third component in the first equation of (5.1) is just the "hydrostatic balance" $\partial_3 p_t = m_t$, see [34, 81].

For this reason, in the next section, we start considering the technically simpler case of the 2-dimentional periodic setting:

$$\begin{cases} \partial_t u_t^g + \left(u_t \cdot \nabla\right)u_t^g + \nabla p_t = -J u_t & \text{in } \mathbb{T}^2 \times (0, \infty) \\ u_t^g(x) = J\nabla p_t(x) & \text{in } \mathbb{T}^2 \times (0, \infty) \\ \nabla \cdot u_t(x) = 0 & \text{in } \mathbb{T}^2 \times (0, \infty) \\ p_0 = p^0 & \text{in } \mathbb{T}^2. \end{cases} \quad (5.2)$$

This time $J$ is the $\pi/2$-rotation matrix given by

$$J := \begin{pmatrix} 0 & -1 \\ 1 & 0 \end{pmatrix}.$$

Let us now focus for a moment on the 3-dimensional case and on some formal computation. If we introduce the "geopotential" (see [81])

$$P_t := p_t + \frac{1}{2}(x_1^2 + x_2^2), \quad (5.3)$$

equation (5.1), can be rewritten as

$$\begin{cases} \partial_t \nabla P_t + (u_t \cdot \nabla)\nabla P_t = J(\nabla P_t - x) \\ \nabla \cdot u_t = 0 \\ u_t \cdot \nu_\Omega = 0 \\ P_0 = p^0 + \frac{1}{2}(x_1^2 + x_2^2). \end{cases} \quad (5.4)$$

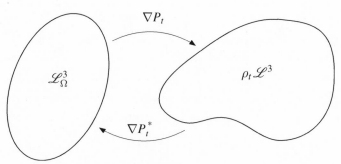

**Figure 5.1.** The dual change of coordinates.

Cullen stability principle (see [34, Section 3.2]) asserts that for a physically admissible solution $P_t$ should be a convex function for every time.

Now consider the measure[3] $\rho_t = (\nabla P_t)_\sharp \mathscr{L}^3_\Omega$, then for any test function $\varphi \in C^1_c$ we compute

$$\frac{d}{dt} \int \varphi(x) \, d\rho_t(x) = \frac{d}{dt} \int_\Omega \varphi(\nabla P_t(y)) \, dy$$

$$= \int_\Omega \nabla\varphi(\nabla P_t(y)) \cdot \frac{d}{dt} \nabla P_t(y) \, dy$$

$$= -\int_\Omega \nabla\varphi(\nabla P_t(y))$$

$$\cdot \left\{ u_t(y) D^2 P_t(y) - J(\nabla P_t(y) - y) \right\} dy$$

$$= -\int_\Omega \nabla\big[\varphi(\nabla P_t(y))\big] \cdot u_t(y) \, dy$$

$$+ \int_\Omega \nabla\varphi(\nabla P_t(y)) \cdot J(\nabla P_t(y) - y) \, dy$$

$$= \int \nabla\varphi(x) \cdot J(x - \nabla P_t^*(x)) \, d\rho_t(x).$$

---

[2] We are using the notation $f_t$ to denote the function $f(t, \cdot)$.

[3] $\mathscr{L}^3_\Omega$ denotes the *normalized* Lebesgue measure on $\Omega$:

$$\frac{1}{|\Omega|} \mathscr{L}^3 \llcorner \Omega,$$

in particular we can (and do) assume without loss of generality that $|\Omega| = 1$.

In other words, $\rho_t$ solves the following "dual" equation:

$$\begin{cases} \partial_t \rho_t + \nabla \cdot (\mathcal{U}_t \rho_t) = 0 \\ \mathcal{U}_t(x) = J(x - \nabla P_t^*(x)) \\ \rho_t = (\nabla P_t)_\sharp \mathscr{L}_\Omega^3 \end{cases} \tag{5.5}$$

Here $P_t^*$ is the convex conjugate of $P_t$, namely

$$P_t^*(y) := \sup_{x \in \Omega}(y \cdot x - P_t(x)) \qquad \forall y \in \mathbb{R}^3.$$

Let us rewrite $\rho_t = (\nabla P_t)_\sharp \mathscr{L}_\Omega^3$ as $(\nabla P_t)_\sharp^* \rho_t = \mathscr{L}_\Omega^3$. If we now assume that $\Omega$ is *convex*, we know from the discussion in Section 1.2 that $P_t^*$ is an Aleksandrov solution to the Monge-Ampère equation:

$$\det D^2 P_t^* = \rho_t.$$

Thus, we can rewrite (5.5) as

$$\begin{cases} \partial_t \rho_t + \nabla \cdot (\mathcal{U}_t \rho_t) = 0 \\ \mathcal{U}_t(x) = J(x - \nabla P_t^*(x)) \\ \det D^2 P_t^* = \rho_t. \end{cases} \tag{5.6}$$

Notice that the above equation takes the form of a continuity equation where the velocity field is coupled with the density through a time independent elliptic PDE. The most famous equation of this form is, probably, the 2-dimensional incompressible Euler equation in the vorticity formulation:

$$\begin{cases} \partial_t \omega_t + \nabla \cdot (v_t \omega_t) = 0 \\ v_t = J \nabla \psi_t \\ \omega_t = \Delta \psi_t. \end{cases} \tag{5.7}$$

Incidentally we notice here that, in 2 dimension, if we "linearize" (5.6) writing $\rho_t = 1 + \varepsilon \omega_t + o(\varepsilon)$ and $P_t^* = |x|^2/2 + \varepsilon \psi_t + o(\varepsilon)$ and we perform a time scaling $t \to t/\varepsilon$, then, formally, $\omega_t$ and $\psi_t$ solve (5.7), (see [77] for a rigorous discussion).

As we saw the coupling between the velocity field and the density in (5.5) is non-linear, nevertheless thanks to the continuous dependence of optimal transport maps with respect to data it is continuous (see Theorem 1.14). This allows for existence of solutions through, for instance, an explicit Euler scheme. More precisely in [13] Benamou and Brenier (see also [36, 35]) proved the following theorem (actually their theorem is under slightly different assumptions, however, the proof can be adapted verbatim to the case of general probability densities, see the sketch of the proof below)

**Theorem 5.1 (Existence of solutions of** (5.5)**).** *Let* $\Omega$ *be a bounded set* $P_0 : \mathbb{R}^3 \to \mathbb{R}$ *be a convex function such that* $(\nabla P_0)_\sharp \mathscr{L}^3_\Omega \ll \mathscr{L}^3$. *Then there exist convex functions* $P_t, P_t^* : \mathbb{R}^3 \to \mathbb{R}$ *such that* $(\nabla P_t)_\sharp \mathscr{L}^3_\Omega = \rho_t \mathscr{L}^3$, $(\nabla P_t^*)_\sharp \rho_t = \mathscr{L}^3_\Omega$, $\mathcal{U}_t(x) = J(x - \nabla P_t^*(x))$, *and* $\rho_t$ *is a distributional solution to* (5.5), *namely*

$$\iint_{\mathbb{R}^3} \left\{ \partial_t \varphi_t(x) + \nabla \varphi_t(x) \cdot \mathcal{U}_t(x) \right\} \rho_t(x)\, dx\, dt + \int_{\mathbb{R}^3} \varphi_0(x)\rho_0(x)\, dx = 0$$
(5.8)

*for every* $\varphi \in C_c^\infty(\mathbb{R}^3 \times [0, \infty))$.
*Moreover, the following regularity properties hold:*

(i) $\rho_t \mathscr{L}^3 \in C([0, \infty), \mathcal{P}_w(\mathbb{R}^3))$, *where* $\mathcal{P}_w(\mathbb{R}^3)$ *is the space of probability measures endowed with the weak topology induced by the duality with* $C_b(\mathbb{R}^3)$;
(ii) $P_t^* - P_t^*(0) \in L_{\text{loc}}^\infty([0, \infty), W_{\text{loc}}^{1,\infty}(\mathbb{R}^3)) \cap C([0, \infty), W_{\text{loc}}^{1,r}(\mathbb{R}^3))$ *for every* $r \in [1, \infty)$;
(iii) $|\mathcal{U}_t(x)| \leq |x| + \text{diam}(\Omega)$ *for almost every* $x \in \mathbb{R}^3$, *for all* $t \geq 0$.

*Sketch of the proof of Theorem* 5.1. Say we want to prove the existence of a solution of (5.5) up to time $T = 1$. To do this we fix $h$ small such that $1/h \in \mathbb{N}$, and divide $[0, 1]$ in intervals of length $h$:

$$[0, 1] = \bigcup_{k=1}^{1/h} [(k-1)h, kh].$$

We define approximate solutions $(\nabla P_t^*)^h$ and $\rho_t^h$ inductively as follows: Suppose they are defined up to time $t \leq (k-1)h$, then for $t \in ((k-1)h, kh]$

$$(\nabla P_t^*)^h := (\nabla P_{kh}^*)^h \quad \text{where} \quad \left[ (\nabla P_{kh}^*)^h \right]_\sharp \rho_{(k-1)h}^h = \mathscr{L}^3_\Omega,$$

$$\rho_t^h \quad \text{such that} \quad \begin{cases} \partial_t \rho_t^h + \nabla \cdot (\rho_t^h \mathcal{U}_t^h) = 0 \\ \mathcal{U}_t^h = J\left(x - (\nabla P_t^*)^h\right) \\ \rho^h(h(k-1), \cdot) = \rho_{(k-1)h}^h. \end{cases}$$

In other words, for $t \in [(k-1)h, kh]$, we define the function $(\nabla P_t^*)^h$ to be identically equal to the optimal transport map between $\rho_{(k-1)h}^h$ and $\mathscr{L}^3_\Omega$ and, once constructed the (constant in time) velocity field in the natural way, we let evolve the density $\rho_{(k-1)h}^h$ according for time $h$ according to the continuity equation. Notice that we are not assuming that $\rho$ has finite second moment, nevertheless Theorem 1.13 ensures the existence of an "optimal map" $(\nabla P_{kh}^*)^h$. We remark that to be sure about existence of well behaved solutions of the continuity equation, one has to rely on the Ambrosio-Di Perna-Lions theory of Regular Lagrangian Flow (see [2, 3, 46]). Another possibility, which is the one used in [13], is to regularize all the data in an appropriate way. Finally, passing to the limit as $h$ goes to 0 we obtain a solution to (5.5), see the above mentioned papers for details.                                                                           $\square$

Observe that, by Theorem 5.1(ii), $t \mapsto \rho_t \mathscr{L}^3$ is weakly continuous, so $\rho_t$ is a well-defined function *for every* $t \geq 0$.

Up to now we have rigorously proven the existence of a (distributional solution) associated to the dual equation (5.5). It is possible from this to recover a solution to (5.1)? Given a solution $(\rho_t, P_t^*)$ of (5.5), we can construct the convex conjugate $P_t$ of $P_t^*$. Then an easy (formal) computation shows that the couple $(p_t, u_t)$ defined by

$$\begin{cases} p_t(x) := P_t(x) - |x|^2/2 \\ u_t(x) := [\partial_t \nabla P_t^*](\nabla P_t(x)) + [D^2 P_t^*](\nabla P_t(x)) J(\nabla P_t(x) - x) \end{cases} \tag{5.9}$$

solves (5.1). To make this computation rigorous we have to give a meaning to the velocity field. Indeed, without any assumption on the domain $\Omega$, $P_t^*$ is just a convex function and its (distributional) Hessian is merely a matrix valued measure (see Appendix A). Moreover we need also to understand the regularity of the term $[\partial_t \nabla P_t^*](\nabla P_t(x))$. This will be done in the subsequent sections thanks to the Sobolev regularity of optimal transport maps established in Chapter 3.

## 5.2. The 2-dimensional periodic case

In this section we establish the first rigorous result about existence of distributional solution of (5.2) in the 2-dimensional periodic case.[4] Namely we prove that the velocity field $u_t$ defined in (5.9) is a well defined $L^1$ function and that the couple $(p_t, u_t)$ is a distributional solution of (5.2). Finally, in the last part of the Section, we also show that, although $u_t$ is merely a summable vector field, associated to it there is a natural notion of measure preserving flow, thus recovering the result of Cullen and Feldman [35] on the existence of Lagrangian solutions to the Semigeostrophic Equations in physical space.

---

[4] See however [77] , where a small time existence result for periodic smooth solutions of (5.5) is proven: it is clear from the proof below that the smooth solutions of the dual equation can be transformed in solutions to the original Equation (5.2)

We now want to define the notion of distributional solution to the semi-geostoprophic equations. Notice that in the 2-periodic setting we cannot give a distributional meaning to (5.4), indeed the function $\nabla P_t$ should be thought as map from the torus into itself, namely $\nabla P_t = \exp_x(\nabla p_t)$ (see Theorem 5.4 below). Since there is no natural notion of duality with test functions for maps with value in a manifold, we prefer to write an equation for *vector fields* for which is easier to give a weak meaning. Notice, instead, that the notion of distributional solution to (5.5) introduced in Theorem 5.1 makes perfectly sense on the torus since the term $x - \nabla P_t^*$ is actually a vector field, namely the gradient of the map $p_t^*$, the $d_{\mathbb{T}^2}^2$-conjugate of $p_t$, see Section 1.3 and Theorem 5.4 below. Moreover, thanks to Theorem 5.4, the proof of the existence of a (periodic) solution to (5.5) is exactly the same of Theorem 5.1.

Substituting the relation $\boldsymbol{u}_t^g = J\nabla p_t$ into the equation, the system (5.2) can be rewritten as

$$
\begin{cases}
\partial_t J \nabla p_t + J D^2 p_t \boldsymbol{u}_t + \nabla p_t + J \boldsymbol{u}_t = 0 \\
\nabla \cdot \boldsymbol{u}_t = 0 \\
p_0 = p^0
\end{cases}
\tag{5.10}
$$

with $\boldsymbol{u}_t$ and $p_t$ periodic.

Thus the natural notion of distributional solution is the following:

**Definition 5.2.** Let $p : \mathbb{T}^2 \times (0, \infty) \to \mathbb{R}$ and $u : \mathbb{T}^2 \times (0, \infty) \to \mathbb{R}^2$. We say that $(p, \boldsymbol{u})$ is a *weak Eulerian solution* of (5.10) if:

- $|\boldsymbol{u}| \in L^\infty((0, \infty), L^1(\mathbb{T}^2))$, $p \in L^\infty((0, \infty), W^{1,\infty}(\mathbb{T}^2))$, and $p_t(x) + |x|^2/2$ is convex for any $t \geq 0$.
- For every $\phi \in C_c^\infty(\mathbb{T}^2 \times [0, \infty))$, it holds

$$
\int_0^\infty \int_{\mathbb{T}^2} J\nabla p_t(x)\Big\{\partial_t \phi_t(x) + \boldsymbol{u}_t(x) \cdot \nabla \phi_t(x)\Big\}
$$

$$
- \Big\{\nabla p_t(x) + J\boldsymbol{u}_t(x)\Big\}\phi_t(x)\, dx\, dt
\tag{5.11}
$$

$$
+ \int_{\mathbb{T}^2} J\nabla p_0(x)\phi_0(x)\, dx = 0.
$$

- For a.e. $t \in (0, \infty)$ it holds

$$
\int_{\mathbb{T}^2} \nabla\psi(x) \cdot \boldsymbol{u}_t(x)\, dx = 0 \qquad \text{for all } \psi \in C^\infty(\mathbb{T}^2).
\tag{5.12}
$$

We can now state our main result.

**Theorem 5.3.** *Let $p_0$ : $\mathbb{R}^2 \to \mathbb{R}$ be a $\mathbb{Z}^2$-periodic function such that $p_0(x) + |x|^2/2$ is convex, and assume that the measure $(\mathrm{Id} + \nabla p_0)_{\sharp} \mathscr{L}^2$ is absolutely continuous with respect to $\mathscr{L}^2$ with density $\rho_0$, namely*

$$(\mathrm{Id} + \nabla p_0)_{\sharp} \mathscr{L}^2 = \rho_0 \mathscr{L}^2.$$

*Moreover, let us assume that both $\rho_0$ and $1/\rho_0$ belong to $L^\infty(\mathbb{R}^2)$.*

*Let $\rho_t$ be the solution of (5.5) given by Theorem 5.1 and let $P_t$ : $\mathbb{R}^2 \to \mathbb{R}$ be the (unique up to an additive constant) convex function such that $(\nabla P_t)_{\sharp} \mathscr{L}^2 = \rho_t \mathscr{L}^2$ and $P_t(x) - |x|^2/2$ is $\mathbb{Z}^2$-periodic (see Theorem 5.4 below), $P_t^*$ : $\mathbb{R}^2 \to \mathbb{R}$ its convex conjugate.*

*Then the couple $(p_t, \boldsymbol{u}_t)$ defined in (5.9) is a weak Eulerian solution of (5.10), in the sense of Definition 5.2.*

Before starting the proof of the above Theorem, we recall the following key theorem, due to Cordero-Erausquin [32] about existence (and regularity) of optimal transport maps for periodic measures. It is actually a corollary of the more general Theorem 1.29, but it can be also proven directly.

**Theorem 5.4 (Existence of optimal maps on $\mathbb{T}^2$).** *Let $\mu$ and $\nu$ be $\mathbb{Z}^2$-periodic Radon measures on $\mathbb{R}^2$ such that $\mu([0, 1)^2) = \nu([0, 1)^2) = 1$ and let $\mu = \rho \mathscr{L}^2$ with $\rho > 0$ almost everywhere. Then there exists a unique (up to an additive constant) convex function $P$ : $\mathbb{R}^2 \to \mathbb{R}$ such that $(\nabla P)_{\sharp} \mu = \nu$ and $P - |x|^2/2$ is $\mathbb{Z}^2$-periodic. Moreover*

$$\nabla P(x + h) = \nabla P(x) + h \qquad \text{for a.e. } x \in \mathbb{R}^2, \ \forall h \in \mathbb{Z}^2, \qquad (5.13)$$

$$|\nabla P(x) - x| \le \mathrm{diam}(\mathbb{T}^2) = \frac{\sqrt{2}}{2} \qquad \text{for a.e. } x \in \mathbb{R}^2. \qquad (5.14)$$

*In addition, if $\mu = \rho \mathscr{L}^2$, $\nu = \sigma \mathscr{L}^2$, and there exists a constant $0 < \lambda \le 1$ such that $\lambda \le \rho, \sigma \le 1/\lambda$, then $P$ is a strictly convex Aleksandrov solution of*

$$\det D^2 P(x) = f(x), \qquad \text{with } f(x) = \frac{\rho(x)}{\sigma(\nabla P(x))}.$$

*Proof.* By Theorem 1.29 we know the existence of a unique transport map $T = \exp(\nabla \tilde{p})$ for some $d_{T^2}^2$-convex function $\tilde{p}$. Since the exponential map on the torus is given by

$$\exp_x(v) = x + v \quad \mod \mathbb{Z}^2,$$

to prove the first part we only have to show that (identifying periodic functions with functions defined on the torus)

$$\left. \begin{array}{ll} P & \text{convex} \\ p := P - |x|^2/2 & \text{periodic} \end{array} \right\} \quad \Longleftrightarrow \quad p \quad d_{T^2}^2\text{-convex}.$$

To see this, observe that, under our assumption, also $p^*(y) := P^*(y) - |y|^2/2$ is $\mathbb{Z}^2$-periodic. Hence, since

$$P(x) = \sup_{y \in \mathbb{R}^2} x \cdot y - P^*(y),$$

we get that the function $p(x) = P(x) - |x|^2/2$ satisfies

$$\begin{aligned} p(x) &= \sup_{y \in \mathbb{R}^2} \left( - \frac{|y - x|^2}{2} - P^*(y) + \frac{|y|^2}{2} \right) \\ &= \sup_{y \in [0,1]^2} \sup_{h \in \mathbb{Z}^2} \left( - \frac{|y + h - x|^2}{2} - p^*(y + h) \right) \\ &= \sup_{y \in T^2} \left( - \frac{d_{T^2}^2(x, y)}{2} - p^*(y) \right), \end{aligned}$$

where we used that $p^*(y)$ is $\mathbb{Z}^2$-periodic and that the geodesic distance on the flat torus is given by

$$d_{T^2}(x, y) = \inf_{h \in \mathbb{Z}^2} |y - x + h|.$$

This proves the claim and that $p^*$ is its $d_{T^2}^2$-conjugate. The fact the $P_t$ is a Aleksandrov solution to the Monge-Ampère equation follows by the arguments of Section 1.2.                                              $\square$

Since $P$ is an Aleksandrov solution to the Monge Ampere equation, the results in Chapters 2 and 3 apply, yielding the following:

**Theorem 5.5 (Space regularity of optimal maps on $T^2$).** *Let $\mu = \rho \mathscr{L}^2$, $\nu = \sigma \mathscr{L}^2$, and let $P$ be as in Theorem 5.4 with $\int_{T^2} P\, dx = 0$. Then:*

(i) *$P \in C^{1,\beta}(T^2)$ for some $\beta = \beta(\lambda) \in (0, 1)$, and there exists a constant $C = C(\lambda)$ such that*

$$\|P\|_{C^{1,\beta}} \le C.$$

(ii) *$P \in W^{2,1}(T^2)$, more precisely there exist a constant $C = C(\lambda)$ and an exponent $\gamma_0 = \gamma_0(\lambda) > 1$ such that*

$$\int_{T^2} |D^2 P|^{\gamma_0}\, dx \le C.$$

(iii) *If* $\rho$, $\sigma$ $\in$ $C^{k,\alpha}(\mathbb{T}^2)$ *for some* $k \in \mathbb{N}$ *and* $\alpha \in (0,1)$, *then* $P \in$ $C^{k+2,\alpha}(\mathbb{T}^2)$ *and there exists a constant* $C = C(\lambda, \|\rho\|_{C^{k,\alpha}}, \|\sigma\|_{C^{k,\alpha}})$ *such that*

$$\|P\|_{C^{k+2,\alpha}} \leq C.$$

*Moreover, there exist two positive constants* $c_1$ *and* $c_2$, *depending only on* $\lambda$, $\|\rho\|_{C^{0,\alpha}}$, *and* $\|\sigma\|_{C^{0,\alpha}}$, *such that*

$$c_1 \operatorname{Id} \leq D^2 P(x) \leq c_2 \operatorname{Id} \qquad \forall x \in \mathbb{T}^2.$$

### 5.2.1. The regularity of the velocity field

The following proposition, which provides the Sobolev regularity of $t \mapsto \nabla P_t^*$, is our main technical tool. Notice that, in order to prove Theorem 5.3, only finiteness of the left hand side in (5.15) would be needed, and the proof of this fact involves only a smoothing argument, the Sobolev regularity estimates of Chapter 3, collected in Theorem 5.5(ii), and the argument of [76, Theorem 5.1]. However, the continuity of transport map in the strong Sobolev topology proved in Chapter 4 allows to show the validity of the natural *a priori* estimate on the left hand side in (5.15).

**Proposition 5.6 (Time regularity of optimal maps).** *Let* $\rho_t$ *and* $P_t$ *be as in Theorem* 5.1. *Then* $\nabla P_t^* \in W^{1,1}_{\mathrm{loc}}(\mathbb{T}^2 \times [0,\infty); \mathbb{R}^2)$, *namely there exist constants* $C(\lambda)$ $\gamma_0 = \gamma_0(\lambda)$ *such that, for almost every* $t \geq 0$,

$$\int_{\mathbb{T}^2} \rho_t |\partial_t \nabla P_t^*|^{\frac{2\gamma_0}{1+\gamma_0}} dx$$
$$\leq C(\lambda)\left( \int_{\mathbb{T}^2} \rho_t |D^2 P_t^*|^{\gamma_0} dx + \operatorname{ess\,sup}_{\mathbb{T}^2} (\rho_t |\mathcal{U}_t|^2) \int_{\mathbb{T}^2} |D^2 P_t^*| dx \right). \tag{5.15}$$

To prove the above proposition we need to understand, in the smooth setting, what is the regularity of the map $t \mapsto \nabla P_t$.

**Lemma 5.7 (Space-time regularity of transport).** *Let* $k \in \mathbb{N} \cup \{0\}$, *and let* $\rho \in C^\infty(\mathbb{T}^2 \times [0,\infty))$ *and* $U \in C^\infty(\mathbb{T}^2 \times [0,\infty); \mathbb{R}^2)$ *satisfy*

$$0 < \lambda \leq \rho_t(x) \leq 1/\lambda < \infty \qquad \forall (x,t) \in \mathbb{T}^2 \times [0,\infty),$$
$$\partial_t \rho_t + \nabla \cdot (\mathcal{U}_t \rho_t) = 0 \qquad in\ \mathbb{T}^2 \times [0,\infty),$$

*and* $\int_{\mathbb{T}^2} \rho_t\, dx = 1$ *for all* $t \geq 0$. *Let us consider convex conjugate maps* $P_t$ *and* $P_t^*$ *such that* $P_t(x) - |x|^2/2$ *and* $P_t^*(y) - |y|^2/2$ *are* $\mathbb{Z}^2$-*periodic,* $(\nabla P_t^*)_\sharp \rho_t = \mathscr{L}^2_{\mathbb{T}^2}$, $(\nabla P_t)_\sharp \mathscr{L}^2_{\mathbb{T}^2} = \rho_t$. *Then:*

(i) $P_t^* - \fint_{\mathbb{T}^2} P_t^* \in \operatorname{Lip}_{\mathrm{loc}}([0,\infty); C^k(\mathbb{T}^2))$ *for any* $k \in \mathbb{N}$.

(ii) *The following linearized Monge-Ampère equation holds:*

$$\nabla \cdot \left(\rho_t (D^2 P_t^*)^{-1} \partial_t \nabla P_t^*\right) = -\nabla \cdot (\rho_t \mathcal{U}_t). \tag{5.16}$$

*Proof.* Let us fix $T > 0$. From the regularity theory for the Monge-Ampère equation (see Theorem 5.5) we obtain that $P_t \in C^\infty(\mathbb{R}^2)$, uniformly for $t \in [0, T]$, and there exist universal constants $c_1, c_2 > 0$ such that

$$c_1 \operatorname{Id} \le D^2 P_t^*(x) \le c_2 \operatorname{Id} \qquad \forall (x, t) \in \mathbb{T}^2 \times [0, T]. \tag{5.17}$$

Since $\nabla P_t^*$ is the inverse of $\nabla P_t$, by the smoothness of $P_t$ and (5.17) we deduce that $P_t^* \in C^\infty(\mathbb{R}^2)$, uniformly on $[0, T]$.

Now, to prove (i), we need to investigate the time regularity of $P_t^* - \fint_{\mathbb{T}^2} P_t^*$. Moreover, up to adding a time dependent constant to $P_t$, we can assume without loss of generality that $\int_{\mathbb{T}^2} P_t^* = 0$ for all $t$. By the condition $(\nabla P_t^*)_\sharp \rho_t = \mathscr{L}_{\mathbb{T}^2}^2$ we get that for any $0 \le s, t \le T$ and $x \in \mathbb{R}^2$ it holds

$$\frac{\rho_s(x) - \rho_t(x)}{s - t} = \frac{\det(D^2 P_s^*(x)) - \det(D^2 P_t^*(x))}{s - t}$$

$$= \sum_{i,j=1}^{2} \left( \int_0^1 \frac{\partial \det}{\partial \xi_{ij}} (\tau D^2 P_s^*(x) + (1 - \tau) D^2 P_t^*(x)) \, d\tau \right)$$

$$\times \frac{\partial_{ij} P_s^*(x) - \partial_{ij} P_t^*(x)}{s - t}.$$

$$\tag{5.18}$$

Recall that, given a symmetric invertible matrix $A \in \mathbb{R}^{2 \times 2}$,

$$\frac{\partial \det(A)}{\partial \xi_{ij}} = M_{ij}(A), \tag{5.19}$$

where $M(A)$ is the cofactor matrix of $A$, *i.e.* the one which satisfies the identity

$$M(A) = \det(A) \, A^{-1}. \tag{5.20}$$

Moreover, if $A$ satisfies $c_1 \operatorname{Id} \le A \le c_2 \operatorname{Id}$ for some positive constants $c_1, c_2$, then

$$c_1 \operatorname{Id} \le M(A) \le c_2 \operatorname{Id}. \tag{5.21}$$

Hence, from (5.18), (5.19), (5.17), and (5.21), it follows that

$$\frac{\rho_s - \rho_t}{s - t} = \sum_{i,j=1}^{2} \left( \int_0^1 M_{ij}(\tau D^2 P_s^* + (1 - \tau) D^2 P_t^*) \, d\tau \right) \partial_{ij} \left( \frac{P_s^* - P_t^*}{s - t} \right),$$

$$\tag{5.22}$$

with

$$c_1 \operatorname{Id} \le \int_0^1 M_{ij}(\tau D^2 P_s^* + (1 - \tau)D^2 P_t^*)\, d\tau \le c_2 \operatorname{Id}$$

Since $D^2 P_t^*$ is smooth in space, uniformly on $[0, T]$, by classical elliptic regularity theory[5] it follows that for any $k \in \mathbb{N}$ and $\alpha \in (0, 1)$ there exists a constant $C := C(\|(\rho_s - \rho_t)/(s - t)\|_{C^{k,\alpha}(\mathbb{T}^2 \times [0,T])})$ such that

$$\left\| \frac{P_s^*(x) - P_t^*(x)}{s - t} \right\|_{C^{k+2,\alpha}(\mathbb{T}^2)} \le C.$$

This proves point (i) in the statement. To prove the second part, we let $s \to t$ in (5.22) to obtain

$$\partial_t \rho_t = \sum_{i,j=1}^2 M_{ij}(D^2 P_t^*(x))\, \partial_t \partial_{ij} P_t^*(x). \tag{5.23}$$

Taking into account the continuity equation and the well-known divergence-free property of the cofactor matrix

$$\sum_i \partial_i M_{ij}(D^2 P_t^{*}(x)) = 0, \qquad j = 1, 2,$$

we can rewrite (5.23) as

$$-\nabla \cdot (\mathcal{U}_t \rho_t) = \sum_{i,j=1}^2 \partial_i \big( M_{ij}(D^2 P_t^*(x))\, \partial_t \partial_j P_t^*(x) \big).$$

Hence, using (5.20) and the Monge-Ampère equation $\det(D^2 P_t^*) = \rho_t$, we finally get (5.16).    $\square$

*Proof of Proposition* 5.6. We closely follow the proof of [76, Theorem 5.1], and we split the proof in two parts. In the first step we assume that

$$\rho_t \in C^\infty(\mathbb{T}^2 \times \mathbb{R}), \ \mathcal{U}_t \in C^\infty(\mathbb{T}^2 \times \mathbb{R}; \mathbb{R}^2), \tag{5.24}$$

$$0 < \lambda \le \rho_t \le 1/\lambda < \infty, \tag{5.25}$$

$$\partial_t \rho_t + \nabla \cdot (\mathcal{U}_t \rho_t) = 0, \tag{5.26}$$

$$(\nabla P_t)_\sharp \mathscr{L}^2_{\mathbb{T}^2} = \rho_t \mathscr{L}^2_{\mathbb{T}^2}, \tag{5.27}$$

and we prove that (5.15) holds for every $t \ge 0$. In the second step we prove the general case through an approximation argument.

---

[5] Note that equation (5.18) is well defined on $\mathbb{T}^2$ since $P_t^* - P_s^*$ is $\mathbb{Z}^2$-periodic. We also observe that $P_t^* - P_s^*$ has average zero on $\mathbb{T}^2$.

*Step* 1 : *The regular case.* Let us assume that the regularity assumptions (5.24), (5.25), (5.26), (5.27) hold. Moreover, up to adding a time dependent constant to $P_t$, we can assume without loss of generality that $\int_{\mathbb{T}^2} P_t^* = 0$ for all $t \geq 0$, so that by Lemma 5.7 we have $\partial_t P_t^* \in C^\infty(\mathbb{T}^2)$. Fix $t \geq 0$. Multiplying (5.16) by $\partial_t P_t^*$ and integrating by parts, we get

$$
\int_{\mathbb{T}^2} \rho_t |(D^2 P_t^*)^{-1/2} \partial_t \nabla P_t^*|^2 \, dx = \int_{\mathbb{T}^2} \rho_t \partial_t \nabla P_t^* \cdot (D^2 P_t^*)^{-1} \partial_t \nabla P_t^* \, dx
$$
$$
= - \int_{\mathbb{T}^2} \rho_t \partial_t \nabla P_t^* \cdot \mathcal{U}_t \, dx.
$$
(5.28)

(Since the matrix $D^2 P_t^*(x)$ is nonnegative, both its square root and the square root of its inverse are well-defined.) From Cauchy-Schwartz inequality it follows that the right-hand side of (5.28) can be rewritten and estimated with

$$
- \int_{\mathbb{T}^2} \rho_t \partial_t \nabla P_t^* \cdot (D^2 P_t^*)^{-1/2} (D^2 P_t^*)^{1/2} \mathcal{U}_t \, dx
$$
$$
\leq \left( \int_{\mathbb{T}^2} \rho_t |(D^2 P_t^*)^{-1/2} \partial_t \nabla P_t^*|^2 \, dx \right)^{1/2} \left( \int_{\mathbb{T}^2} \rho_t |(D^2 P_t^*)^{1/2} \mathcal{U}_t|^2 \, dx \right)^{1/2}.
$$
(5.29)

Moreover, the second factor in the right-hand side of (5.29) can be estimated with

$$
\int_{\mathbb{T}^2} \rho_t \mathcal{U}_t \cdot D^2 P_t^* \mathcal{U}_t \, dx \leq \max_{\mathbb{T}^2} \left( \rho_t |\mathcal{U}_t|^2 \right) \int_{\mathbb{T}^2} |D^2 P_t^*| \, dx.
$$
(5.30)

Hence, from (5.28), (5.29), and (5.30) it follows that

$$
\int_{\mathbb{T}^2} \rho_t |(D^2 P_t^*)^{-1/2} \partial_t \nabla P_t^*|^2 \, dx \leq \max_{\mathbb{T}^2} \left( \rho_t |\mathcal{U}_t|^2 \right) \int_{\mathbb{T}^2} |D^2 P_t^*| \, dx. \quad (5.31)
$$

We now apply Young's inequality

$$
(ab)^{\frac{2\gamma_0}{1+\gamma_0}} \leq C \left( a^{2\gamma_0} + b^2 \right),
$$

with $a = |(D^2 P_t^*)^{1/2}|$ and $b = |(D^2 P_t^*)^{-1/2} \partial_t \nabla P_t^*(x)|$ to deduce the existence of a constant $C$ such that

$$
|\partial_t \nabla P_t^*|^{\frac{2\gamma_0}{1+\gamma_0}} \leq C \left( |(D^2 P_t^*)^{1/2}|^{2\gamma_0} + |(D^2 P_t^*)^{-1/2} \partial_t \nabla P_t^*|^2 \right)
$$
$$
= C \left( |D^2 P_t^*|^{\gamma_0} + |(D^2 P_t^*)^{-1/2} \partial_t \nabla P_t^*|^2 \right).
$$

Integrating the above inequality over $\mathbb{T}^2$ and using (5.31), we finally obtain

$$\int_{\mathbb{T}^2} \rho_t |\partial_t \nabla P_t^*|^{\frac{2\gamma_0}{1+\gamma_0}} \, dx$$

$$\leq C \left( \int_{\mathbb{T}^2} \rho_t |D^2 P_t^*|^{\gamma_0} \, dx + \int_{\mathbb{T}^2} \rho_t |(D^2 P_t^*)^{-1/2} \partial_t \nabla P_t^*|^2 \, dx \right) \quad (5.32)$$

$$\leq C \left( \int_{\mathbb{T}^2} \rho_t |D^2 P_t^*|^{\gamma_0} \, dx + \max_{\mathbb{T}^2} \left( \rho_t |\mathcal{U}_t|^2 \right) \int_{\mathbb{T}^2} |D^2 P_t^*| \, dx \right),$$

which proves (5.15).

*Step* 2 : *The approximation argument.* First of all, we extend the functions $\rho_t$ and $\mathcal{U}_t$ for $t \leq 0$ by setting $\rho_t = \rho_0$ and $\mathcal{U}_t = 0$ for every $t < 0$. We notice that, with this definition, $\rho_t$ solves the continuity equation with velocity $\mathcal{U}_t$ on $\mathbb{R}^2 \times \mathbb{R}$.

Fix now $\sigma_1 \in C_c^\infty(\mathbb{R}^2)$, $\sigma_2 \in C_c^\infty(\mathbb{R})$, define the family of mollifiers $(\sigma^n)_{n \in \mathbb{N}}$ as $\sigma^n(x, t) := n^3 \sigma_1(nx) \sigma_2(nt)$, and set

$$\rho^n := \rho * \sigma^n, \qquad \mathcal{U}^n(x) := \frac{(\rho \mathcal{U}) * \sigma^n}{\rho * \sigma^n}.$$

Since $\lambda \leq \rho \leq 1/\lambda$ then

$$\lambda \leq \rho^n \leq 1/\lambda.$$

Therefore both $\rho^n$ and $\mathcal{U}^n$ are well defined and satisfy (5.24), (5.25), (5.26). Moreover for every $t > 0$ the function $\rho_t^n$ is $\mathbb{Z}^2$-periodic and it is a probability density when restricted to $(0, 1)^2$ (once again we are identifying periodic functions with functions defined on the torus). Let $P_t^n$ be the only convex function such that $(\nabla P_t^n)_\sharp \mathscr{L}_{\mathbb{T}^2}^2 = \rho_t^n$ and its its convex conjugate $P_t^{n*}$ satisfies $\int_{\mathbb{T}^2} P_t^{n*} = 0$ for all $t \geq 0$. Since $\rho_t^n \to \rho_t$ in $L^1(\mathbb{T}^2)$ for any $t > 0$ (recall that, by Theorem 5.1(ii), $\rho_t$ is weakly continuous in time), from Theorem 1.14 it follows that

$$\nabla P_t^{n*} \to \nabla P_t^* \qquad \text{in } L^1(\mathbb{T}^2) \quad (5.33)$$

for any $t > 0$. Moreover, by Theorems 5.4 and 5.5(ii), there exist constants $C$, $\gamma_0$ depending only on $\lambda$ such that

$$\int_{\mathbb{T}^2} \rho_t^n |D^2 P_t^{n*}|^{\gamma_0} \, dx \leq C,$$

and by Theorem 4.1, it follows that (taking $\gamma_0$ slightly smaller than the optimal one):

$$\int_{\mathbb{T}^2} \rho_t^n |D^2 P_t^{n*}|^{\gamma_0} \, dx \rightarrow \int_{\mathbb{T}^2} \rho_t |D^2 P_t^*|^{\gamma_0} \, dx, \qquad (5.34)$$

$$\int_{\mathbb{T}^2} |D^2 P_t^{n*}| \, dx \rightarrow \int_{\mathbb{T}^2} |D^2 P_t^*| \, dx. \qquad (5.35)$$

Finally, since the function $(w, t) \mapsto F(w, t) = |w|^2/t$ is convex on $\mathbb{R}^2 \times (0, \infty)$, by Jensen inequality we get

$$\|\rho^n |\mathcal{U}^n|^2\|_\infty = \|F(\rho^n \mathcal{U}^n, \rho^n)\|_\infty \leq \|\rho |\mathcal{U}|^2\|_\infty. \qquad (5.36)$$

It is clear that $\rho_t^n \partial_t \nabla P_t^{n*}$ weakly converge in $L^1(\mathbb{T}^2 \times [0, T])$ to $\rho_t \partial_t \nabla P_t^*$ for every $T > 0$. Taking (5.34), (5.35), (5.36) and (5.37) into account, and applying Step1 to $\rho_t^n, \mathcal{U}_t^n$ we deduce that, for every test functions $\phi(t), \varphi(x)$ with $\phi$ positive, $\int_{\mathbb{T}^2} |\varphi|^{\frac{2\gamma_0}{\gamma_0 - 1}} dx \leq 1$, [6]

$$\int_0^T \int_{\mathbb{T}^2} \phi(t) \varphi(x) \rho_t \partial_t \nabla P_t^* dx dt$$

$$= \lim_{n \to \infty} \int_0^T \int_{\mathbb{T}^2} \phi(t) \varphi(x) \rho_t^n \partial_t \nabla P_t^{n*} \, dx dt$$

$$\leq C(\lambda) \liminf_{n \to \infty} \int_0^T \phi(t) \int_{\mathbb{T}^2} \rho_t^n |\partial_t \nabla P_t^{n*}|^{\frac{2\gamma_0}{1 + 2\gamma_0}} \, dx dt$$

$$\leq C(\lambda) \liminf_{n \to \infty} \int_0^T \phi(t)$$

$$\times \left( \int_{\mathbb{T}^2} \rho_t^n |D^2 P_t^{n*}|^{\gamma_0} dx + \max_{\mathbb{T}^2} \left( \rho_t^n |\mathcal{U}_t^n|^2 \right) \int_{\mathbb{T}^2} |D^2 P_t^{n*}| \, dx \right) dt$$

$$= C(\lambda) \int_0^T \phi(t)$$

$$\times \left( \int_{\mathbb{T}^2} \rho_t |D^2 P_t^*|^{\gamma_0} \, dx + \max_{\mathbb{T}^2} \left( \rho_t |\mathcal{U}_t|^2 \right) \int_{\mathbb{T}^2} |D^2 P_t^*| \, dx \right) dt.$$

$$(5.37)$$

---

[6] Notice that, since $\lambda \leq \rho_t^n \leq 1/\lambda$, all the Lebesgue spaces $L^p(\rho_t^n dx)$, $L^p(\rho_t dx)$ are isomorophic to $L^p(dx)$ with constants depending only on $\lambda$.

An easy density argument based on the separability of $L^{\frac{2\gamma_0}{\gamma_0-1}}$ thus implies that, for almost every $t \in [0, T]$,

$$\int_{\mathbb{T}^2} \varphi \rho_t \partial_t \nabla P_t^* dx$$

$$\leq C(\lambda) \left( \int_{\mathbb{T}^2} \rho_t |D^2 P_t^*|^{\gamma_0} dx + \max_{\mathbb{T}^2} \left( \rho_t |\mathcal{U}_t|^2 \right) \int_{\mathbb{T}^2} |D^2 P_t^*| dx \right)$$

for all $\varphi$ such that $\int_{\mathbb{T}^2} |\varphi|^{\frac{2\gamma_0}{\gamma_0-1}} dx \leq 1$, from which the desired result follows (see the footnote at the end of the page). $\qquad\square$

It is clear from the proof of Proposition 5.6 that the particular coupling between the velocity field $\mathcal{U}_t$ and the transport map $P_t$ is not used. Actually, using Theorem 5.5(ii), and arguing again as in the proof of [76, Theorem 5.1], the following more general statement holds (compare with [76, Theorem 5.1, Equations (27) and (29)]):

**Proposition 5.8.** *Let $\rho_t$ and $v_t$ be such that $0 < \lambda \leq \rho_t \leq 1/\lambda < \infty$, $v_t \in L^{\infty}_{\text{loc}}(\mathbb{T}^2 \times [0, \infty), \mathbb{R}^2)$, and*

$$\partial_t \rho_t + \nabla \cdot (v_t \rho_t) = 0.$$

*Assume that $\int_{\mathbb{T}^2} \rho_t \, dx = 1$ for all $t \geq 0$, let $P_t$ be a convex function such that*

$$(\nabla P_t)_{\sharp} \mathscr{L}^2_{\mathbb{T}^2} = \rho_t \mathscr{L}^2_{\mathbb{T}^2},$$

*and denote by $P_t^*$ its convex conjugate.*

*Then $\nabla P_t$ and $\nabla P_t^*$ belong to $W^{1,1}_{\text{loc}}(\mathbb{T}^2 \times [0, \infty); \mathbb{R}^2)$. Moreover there exists a constants $C(\lambda)$ and $\gamma_0(\lambda) > 1$ such that, for almost every $t \geq 0$,*

$$\int_{\mathbb{T}^2} \rho_t |\partial_t \nabla P_t^*|^{\frac{2\gamma_0}{1+\gamma_0}} dx$$

$$\leq C(\lambda) \left( \int_{\mathbb{T}^2} \rho_t |D^2 P_t^*|^{\gamma_0} dx + \operatorname*{ess\,sup}_{\mathbb{T}^2} \left( \rho_t |v_t|^2 \right) \int_{\mathbb{T}^2} |D^2 P_t^*| dx \right) \quad (5.38)$$

$$\int_{\mathbb{T}^2} |\partial_t \nabla P_t|^{\frac{2\gamma_0}{1+\gamma_0}} dx$$

$$\leq C(\lambda) \left( \int_{\mathbb{T}^2} \rho_t |D^2 P_t|^{\gamma_0} dx + \operatorname*{ess\,sup}_{\mathbb{T}^2} \left( \rho_t |v_t|^2 \right) \int_{\mathbb{T}^2} |D^2 P_t| dx \right). \quad (5.39)$$

*Proof.* We just give a short sketch of the proof. Equation (5.38) can be proved following the same line of the proof of Proposition 5.6. To prove (5.39) notice that by the approximation argument in the second step of the proof of Proposition 5.6 we can assume that the velocity and the density are smooth and hence, arguing as in Lemma 5.7, we have that $P_t$, $P_t^* \in \mathrm{Lip}_{\mathrm{loc}}([0, \infty), C^\infty(\mathbb{T}^2))$. Now, changing variables in the left hand side of (5.31) we get

$$
\int_{\mathbb{T}^2} \left| \left( [D^2 P_t^*](\nabla P_t) \right)^{-1/2} [\partial_t \nabla P_t^*](\nabla P_t) \right|^2 dx
$$

$$
\leq \max_{\mathbb{T}^2} \left( \rho_t |\boldsymbol{v}_t|^2 \right) \int_{\mathbb{T}^2} |D^2 P_t^*| \, dx.
$$

(5.40)

Taking into account the identities

$$
[D^2 P_t^*](\nabla P_t) = \left( D^2 P_t \right)^{-1}
$$

and

$$
[\partial_t \nabla P_t^*](\nabla P_t) + [D^2 P_t^*](\nabla P_t)\partial_t \nabla P_t = 0
$$

which follow differentiating with respect to time and space $\nabla P_t^* \circ \nabla P_t = \mathrm{Id}$, Equation (5.40) becomes

$$
\int_{\mathbb{T}^2} |(D^2 P_t)^{-1/2}\partial_t \nabla P_t|^2 \, dx \leq \max_{\mathbb{T}^2} \left( \rho_t |\boldsymbol{v}_t|^2 \right) \int_{\mathbb{T}^2} |D^2 P_t^*| \, dx.
$$

At this point the proof of (5.39) is obtained arguing as in Proposition 5.6. $\square$

### 5.2.2. Existence of an Eulerian solution

In this section we prove Theorem 5.3. In the proof we will need to test (5.5) with functions which are merely $W^{1,1}$. This is made possible by the following lemma.

**Lemma 5.9.** *Let $\rho_t$ and $P_t$ be as in Theorem* 5.1. *Then* (5.8) *holds for every* $\varphi \in W^{1,1}(\mathbb{T}^2 \times [0, \infty))$ *which is compactly supported in time. (Now $\varphi_0(x)$ has to be understood in the sense of traces.)*

*Proof.* Let $\varphi^n \in C^\infty(\mathbb{T}^2 \times [0, \infty))$ be strongly converging to $\varphi$ in $W^{1,1}$, so that $\varphi_0^n$ converges to $\varphi_0$ in $L^1(\mathbb{T}^2)$. Taking into account that both $\rho_t$ and $\mathcal{U}_t$ are uniformly bounded from above in $\mathbb{T}^2 \times [0, \infty)$, we can apply (5.8) to the test functions $\varphi^n$ and let $n \to \infty$ to obtain the same formula with $\varphi$. $\square$

*Proof of Theorem* 5.3. First of all notice that, thanks to Theorem 5.5(i) and Proposition 5.6, it holds $|D^2 P_t^*|$, $|\partial_t \nabla P_t^*| \in L_{\mathrm{loc}}^\infty([0, \infty), L^1(\mathbb{T}^2))$.

Moreover, since $(\nabla P_t)_\sharp \mathscr{L}^2_{\mathbb{T}^2} = \rho_t \mathscr{L}^2_{\mathbb{T}^2}$, it is immediate to check the function $\boldsymbol{u}$ in (5.9) is well-defined[7] and $|\boldsymbol{u}|$ belongs to $L^\infty_{\mathrm{loc}}([0, \infty), L^1(\mathbb{T}^2))$.

Let $\phi \in C_c^\infty(\mathbb{R}^2 \times [0, \infty))$ be a $\mathbb{Z}^2$-periodic function in space and let us consider the function $\varphi : \mathbb{R}^2 \times [0, \infty) \to \mathbb{R}^2$ given by

$$\varphi_t(y) := J(y - \nabla P_t^*(y))\phi_t(\nabla P_t^*(y)). \tag{5.41}$$

By Theorem 5.4 and the periodicity of $\phi$, $\varphi_t(y)$ is $\mathbb{Z}^2$-periodic in the space variable. Moreover $\varphi_t$ is compactly supported in time, and Proposition 5.6 implies that $\varphi \in W^{1,1}(\mathbb{R}^2 \times [0, \infty))$. So, by Lemma 5.9, each component of the function $\varphi_t(y)$ is an admissible test function for (5.8). For later use, we write down explicitly the derivatives of $\varphi$:

$$\begin{cases} \partial_t \varphi_t(y) = -J[\partial_t \nabla P_t^*](y)\phi_t(\nabla P_t^*(y)) \\ \qquad + J(y - \nabla P_t^*(y))[\partial_t \phi_t](\nabla P_t^*(y)) \\ \qquad + J(y - \nabla P_t^*(y))\big([\nabla \phi_t](P_t^*(y)) \cdot \partial_t \nabla P_t^*(y)\big), \\ \nabla \varphi_t(y) = J(Id - D^2 P_t^*(y))\phi_t(\nabla P_t^*(y)) + J(y - \nabla P_t^*(y)) \\ \qquad \otimes\big([\nabla^T \phi_t](P_t^*(y))D^2 P_t^*(y)\big). \end{cases} \tag{5.42}$$

Taking into account that $(\nabla P_t)_\sharp \mathscr{L}^2_{\mathbb{T}^2} = \rho_t \mathscr{L}^2_{\mathbb{T}^2}$ and that $[\nabla P_t^*](\nabla P_t(x)) = x$ almost everywhere, we can rewrite the boundary term in (5.8) as

$$\begin{aligned} \int_{\mathbb{T}^2} \varphi_0(y)\rho_0(y)\,dy &= \int_{\mathbb{T}^2} J(\nabla P_0(x) - x)\phi_0(x)\,dx \\ &= \int_{\mathbb{T}^2} J\nabla p_0(x)\phi_0(x)\,dx. \end{aligned} \tag{5.43}$$

In the same way, since $\mathcal{U}_t(y) = J(y - \nabla P_t^*(y))$, we can use (5.42) to rewrite the other term as

$$\begin{aligned} &\int_0^\infty \int_{\mathbb{T}^2} \Big\{ \partial_t \varphi_t(y) + \nabla \varphi_t(y) \cdot \mathcal{U}_t(y)\Big\}\rho_t(y)\,dy\,dt \\ &= \int_0^\infty \int_{\mathbb{T}^2} \Big\{ -J[\partial_t \nabla P_t^*](\nabla P_t(x))\phi_t(x) + J(\nabla P_t(x) - x)\partial_t \phi_t(x) \\ &\quad + J(\nabla P_t(x) - x)\big(\nabla \phi_t(x) \cdot [\partial_t \nabla P_t^*](\nabla P_t(x))\big) \\ &\quad + \big[J(Id - D^2 P_t^*(\nabla P_t(x)))\phi_t(x) + J(\nabla P_t(x) - x) \\ &\quad \otimes \big(\nabla^T \phi_t(x)D^2 P_t^*(\nabla P_t(x)))\big]J(\nabla P_t(x) - x)\Big\} \end{aligned} \tag{5.44}$$

---

[7] Note that the composition of $D^2 P_t^*$ with $\nabla P_t$ makes sense. Indeed, by the conditions $(\nabla P_t)_\sharp \mathscr{L}^2_{\mathbb{T}^2} = \rho_t \mathscr{L}^2_{\mathbb{T}^2} \ll \mathscr{L}^2_{\mathbb{T}^2}$, if we change the value of $D^2 P_t^*$ in a set of measure zero, also $[D^2 P_t^*](\nabla P_t)$ will change only on a set of measure zero.

which, taking into account the formula (5.9) for $\boldsymbol{u}$, after rearranging the terms turns out to be equal to

$$
\int_0^\infty \int_{\mathbb{T}^2} \Big\{ J\nabla p_t(x)\big(\partial_t \phi_t(x) + \boldsymbol{u}_t(x) \cdot \nabla \phi_t(x)\big)
$$
$$
+ \big(-\nabla p_t(x) - J\boldsymbol{u}_t(x)\big)\phi_t(x)\Big\}
\tag{5.45}
$$

Hence, combining (5.43), (5.44), (5.45), and (5.8), we obtain the validity of (5.11).

Now we prove (5.12). Given $\phi \in C_c^\infty(0, \infty)$ and a $\mathbb{Z}^2$-periodic function $\psi \in C^\infty(\mathbb{R}^2)$, let us consider the function $\varphi : \mathbb{R}^2 \times [0, \infty) \to \mathbb{R}$ defined by

$$
\varphi_t(y) := \phi(t)\psi(\nabla P_t^*(y)).
\tag{5.46}
$$

As in the previous case, we have that $\varphi$ is $\mathbb{Z}^2$-periodic in the space variable and $\varphi \in W^{1,1}(\mathbb{T}^2 \times [0, \infty))$, so we can use $\varphi$ as a test function in (5.12). Then, identities analogous to (5.42) yield

$$
0 = \int_0^\infty \int_{\mathbb{T}^2} \{\partial_t \varphi_t(y) + \nabla \varphi_t(y) \cdot \boldsymbol{\mathcal{U}}_t(y)\}\, \rho_t(y)\, dy\, dt
$$
$$
= \int_0^\infty \phi'(t) \int_{\mathbb{T}^2} \psi(x)\, dx\, dt
$$
$$
+ \int_0^\infty \phi(t) \int_{\mathbb{T}^2} \Big\{ \nabla \psi(x) \cdot \partial_t \nabla P_t^*(\nabla P_t(x))
$$
$$
+ \nabla^T \psi(x) D^2 P_t^*(\nabla P_t(x)) J(\nabla P_t(x) - x)\Big\}\, dx\, dt
$$
$$
= \int_0^\infty \phi(t) \int_{\mathbb{T}^2} \nabla \psi(x) \cdot \boldsymbol{u}_t(x)\, dx\, dt.
$$

Since $\phi$ is arbitrary we obtain

$$
\int_{\mathbb{T}^2} \nabla \psi(x) \cdot \boldsymbol{u}_t(x)\, dx = 0 \qquad \text{for a.e. } t > 0.
$$

By a standard density argument it follows that the above equation holds outside a negligible set of times independent of the test function $\psi$, thus proving (5.12). □

### 5.2.3. Existence of a Regular Lagrangian Flow for the semigeostrophic velocity field

We start with the definition of Regular Lagrangian Flow for a given vector field $b$, inspired by [2, 3]:

**Definition 5.10.** Given a Borel, locally integrable vector field $b : \mathbb{T}^2 \times (0, \infty) \to \mathbb{R}^2$, we say that a Borel function $F : \mathbb{T}^2 \times [0, \infty) \to \mathbb{T}^2$ is a *Regular Lagrangian Flow (in short RLF) associated to b* if the following two conditions are satisfied.

(a) For almost every $x \in \mathbb{T}^2$ the map $t \mapsto F_t(x)$ is locally absolutely continuous in $[0, \infty)$ and

$$F_t(x) = x + \int_0^t b_s(F_s(x))dx \qquad \forall t \in [0, \infty). \qquad (5.47)$$

(b) For every $t \in [0, \infty)$ it holds $(F_t)_{\#}\mathscr{L}_{\mathbb{T}^2}^2 \leq C\mathscr{L}_{\mathbb{T}^2}^2$, with $C \in [0, \infty)$ independent of $t$.

A particular class of RLFs is the collection of the measure-preserving ones, where (b) is strengthened to

$$(F_t)_{\#}\mathscr{L}_{\mathbb{T}^2}^2 = \mathscr{L}_{\mathbb{T}^2}^2 \qquad \forall t \geq 0.$$

Notice that *a priori* the above definition depends on the choice of the representative of $b$ in the Lebesgue equivalence class, since modifications of $b$ in Lebesgue negligible sets could destroy condition (a). However, a simple argument based on Fubini's theorem shows that the combination of (a) and (b) is indeed invariant (see [2, Section 6]): in other words, if $b = \tilde{b}$ a.e. in $\mathbb{T}^2 \times (0, \infty)$, then every RLF associated to $b$ is also a RLF associated to $\tilde{b}$.

We show existence of a measure-preserving RLF associated to the vector field $u$ defined by

$$u_t(x) = [\partial_t \nabla P_t^*](\nabla P_t(x)) + [D^2 P_t^*](\nabla P_t(x)) J(\nabla P_t(x) - x), \quad (5.48)$$

where $P_t$ and $P_t^*$ are as in Theorem 5.3. Recall also that, under these assumptions, $|u| \in L_{\text{loc}}^\infty([0, \infty), L^1(\mathbb{T}^2))$.

Existence for weaker notion of Lagrangian flow of the semigeostrophic equations was proved by Cullen and Feldman, see [35, Definition 2.4], but since at that time the results of [40] were not available the velocity could not be defined, not even as a function. Hence, they had to adopt a more indirect definition. We shall prove indeed that their flow is a flow according to Definition 5.10.

**Theorem 5.11.** *Let us assume that the hypotheses of Theorem 5.3 are satisfied, and let $P_t$ and $P_t^*$ be the convex functions such that*

$$(\nabla P_t)_{\sharp}\mathscr{L}_{\mathbb{T}^2}^2 = \rho_t \mathscr{L}_{\mathbb{T}^2}^2, \qquad (\nabla P_t^*)_{\sharp}\rho_t \mathscr{L}_{\mathbb{T}^2}^2 = \mathscr{L}_{\mathbb{T}^2}^2.$$

*Then, for $u_t$ given by (5.48) there exists a measure-preserving Regular Lagrangian Flow $F$ associated to $u_t$. Moreover $F$ is invertible in the sense that for all $t \geq 0$ there exist Borel maps $F_t^*$ such that $F_t^*(F_t) = \mathrm{Id}$ and $F_t(F_t^*) = \mathrm{Id}$ a.e. in $\mathbb{T}^2$.*

*Proof.* Let us consider the velocity field in the dual variables $\mathcal{U}_t(x) = J(x - \nabla P_t^*(x))$. Since $P_t^*$ is convex, $\mathcal{U}_t \in BV(\mathbb{T}^2; \mathbb{R}^2)$ uniformly in time (actually, by Theorem 5.5(ii) $\mathcal{U}_t \in W^{1,1}(\mathbb{T}^2; \mathbb{R}^2)$). Moreover $\mathcal{U}_t$ is divergence-free. Hence, by the theory of Regular Lagrangian Flows associated to $BV$ vector fields [2, 3], there exists a unique measure-preserving RLF $G : \mathbb{T}^2 \times [0, \infty) \to \mathbb{T}^2$ associated to $\mathcal{U}$. We remark that the uniqueness of Regular Lagrangian Flows has to be understood in the following way: if $G_1, G_2 : \mathbb{T}^2 \times [0, \infty) \to \mathbb{T}^2$ are two RLFs associated to $U$, then the integral curves $G_1(\cdot, x)$ and $G_2(\cdot, x)$ are equal for $\mathscr{L}^2$-a.e. $x$.

We now define[8]

$$F_t(y) := \nabla P_t^*(G_t(\nabla P_0(y))).  \qquad (5.49)$$

The validity of property (b) in Definition 5.10 and the invertibility of $F$ follow from the same arguments of [35, Propositions 2.14 and 2.17]. Hence we only have to show that property (a) in Definition 5.10 holds.

Let us define $Q^n := B * \sigma^n$, where $B$ is a Sobolev and uniformly continuous extension of $\nabla P^*$ to $\mathbb{T}^2 \times \mathbb{R}$, and $\sigma^n$ is a standard family of mollifiers in $\mathbb{T}^2 \times \mathbb{R}$. It is well known that $Q^n \to \nabla P^*$ locally uniformly and in the strong topology of $W_{\mathrm{loc}}^{1,1}(\mathbb{T}^2 \times [0, \infty))$. Thus, using the measure-preserving property of $G_t$, for all $T > 0$ we get

$$0 = \lim_{n \to \infty} \int_{\mathbb{T}^2} \int_0^T \left\{ |Q_t^n - \nabla P_t^*| + |\partial_t Q_t^n - \partial_t \nabla P_t^*| + |\nabla Q_t^n - D^2 P_t^*| \right\} dy \, dt$$

$$= \lim_{n \to \infty} \int_{\mathbb{T}^2} \int_0^T \left\{ |Q_t^n(G_t) - \nabla P_t^*(G_t)| + |[\partial_t Q_t^n](G_t) - [\partial_t \nabla P_t^*](G_t)| \right.$$

$$\left. + |[\nabla Q_t^n](G_t) - [D^2 P_t^*](G_t)| \right\} dx \, dt.$$

---

[8] Observe that the definition of $F$ makes sense. Indeed, by Theorem 5.5(i), both maps $\nabla P_0$ and $\nabla P_t^*$ are Hölder continuous in space. Morever, by the weak continuity in time of $t \mapsto \rho_t$ (Theorem 5.1(ii)) and the stability results for Aleksandrov solutions of Monge-Ampère, $\nabla P^*$ is continuous both in space and time. Finally, since $(\nabla P_0)_\sharp \mathscr{L}_{\mathbb{T}^2}^2 \ll \mathscr{L}_{\mathbb{T}^2}^2$, if we change the value of $G$ in a set of measure zero, also $F$ will change only on a set of measure zero.

Up to a (not re-labeled) subsequence the previous convergence is pointwise in space, namely, for almost every $x \in \mathbb{T}^2$,

$$\int_0^T \left\{ |Q_t^n(G_t(x)) - \nabla P_t^*(G_t(x))| + |[\partial_t Q_t^n](G_t(x)) - [\partial_t \nabla P_t^*](G_t(x))| \right.$$
$$\left. + |[\nabla Q_t^n](G_t(x)) - [D^2 P_t^*](G_t(x))| \right\} dt \to 0.$$
(5.50)

Hence, since $G$ is a RLF and by assumption

$$(\nabla P_0)_\# \mathscr{L}_{\mathbb{T}^2}^2 \ll \mathscr{L}_{\mathbb{T}^2}^2,$$

for almost every $y$ we have that (5.50) holds at $x = \nabla P_0(y)$, and the function $t \mapsto G_t(x)$ is absolutely continuous on $[0, T]$, with derivative given by

$$\frac{d}{dt} G_t(x) = \mathcal{U}_t(G_t(x)) = J(G_t(x) - \nabla P_t^*(G_t(x))) \quad \text{for a.e. } t \in [0, T].$$

Let us fix such an $y$. Since $Q^n$ is smooth, the function $Q_t^n(G_t(x))$ is absolutely continuous in $[0, T]$ and its time derivative is given by

$$\frac{d}{dt}\left( Q_t^n(G_t(x)) \right) = [\partial_t Q_t^n](G_t(x))$$
$$+ [\nabla Q_t^n](G_t(x)) J(G_t(x) - \nabla P_t^*(G_t(x))).$$

Hence, since $J(G_t(x) - \nabla P_t^*(G_t(x))) = \mathcal{U}(G_t(x))$ is uniformly bounded, from (5.50) we get

$$\lim_{n \to \infty} \frac{d}{dt}\left( Q_t^n(G_t(x)) \right) = [\partial_t \nabla P_t^*](G_t(x))$$
$$+ [D^2 P_t^*](G_t(x)) J(G_t(x) - \nabla P_t^*(G_t(x)))$$
$$:= v_t(y) \quad \text{in } L^1(0, T).$$
(5.51)

Recalling that

$$\lim_{n \to \infty} Q_t^n(G_t(x)) = \nabla P_t^*(G_t(x)) = F_t(y) \qquad \forall t \in [0, T],$$

we infer that $F_t(y)$ is absolutely continuous in $[0, T]$ (being the limit in $W^{1,1}(0, T)$ of absolutely continuous maps). Moreover, by taking the limit as $n \to \infty$ in the identity

$$Q_t^n(G_t(x)) = Q_0^n(G_0(x)) + \int_0^t \frac{d}{d\tau}\left( Q_\tau^n(G_\tau(x)) \right) d\tau,$$

thanks to (5.51) we get

$$F_t(y) = F_0(y) + \int_0^t v_\tau(y)\, d\tau. \tag{5.52}$$

To obtain (5.47) we only need to show that $v_t(y) = u_t(F_t(y))$, which follows at once from (5.48), (5.49), and (5.51).

$\square$

## 5.3. The 3-dimensional case

In this Section we study (5.1) (and its equivalent counterpart (5.10)) in the physical space $\mathbb{R}^3$. The scheme of the proof is the same of the previous section and we will just highlight the main differences through the proofs.

We start noticing the following difficulty in carrying over the strategy of the previous section: if in (5.5) we start with a compactly supported density $\rho_0$, there is no reason for the set $\{\rho_t > 0\}$ to be open. Indeed, a priori, $\mathcal{U}_t$ is only known to be in $BV$ and this is not enough to ensure that $\{\rho_t > 0\}$ is open.

In order to overcome this difficulty we will consider probability measures $\rho_0 = (\nabla P_0)_\sharp \mathcal{L}^3_\Omega$ supported on the whole $\mathbb{R}^3$ and decaying fast enough at infinity. It turns out that these conditions are stable in time along the evolution of (5.5) and allow us to perform a suitable regularization scheme.

We conclude noticing that it would be extremely interesting to consider compactly supported initial data $\rho_0 = (\nabla P_0)_\sharp \mathcal{L}^3_\Omega$. However, overcoming the above mentioned difficulties, seems to require completely new ideas and ingredients.

Let us start giving the definition of distributional solution in this case:

**Definition 5.12.** Let $P : \Omega \times [0, \infty) \to \mathbb{R}$ and $u : \Omega \times [0, \infty) \to \mathbb{R}^3$. We say that $(P, u)$ is a *weak Eulerian solution* of (5.10) if:

- $|u| \in L^\infty_{\text{loc}}((0, \infty), L^1_{\text{loc}}(\Omega))$, $P \in L^\infty_{\text{loc}}((0, \infty), W^{1,\infty}_{\text{loc}}(\Omega))$, and $P_t(x)$ is convex for any $t \geq 0$.
- For every $\phi \in C^\infty_c(\Omega \times [0, \infty))$, it holds

$$\int_0^\infty \int_\Omega \nabla P_t(x) \Big\{ \partial_t \phi_t(x) + u_t(x) \cdot \nabla \phi_t(x) \Big\}$$
$$+ J\Big\{ \nabla P_t(x) - x \Big\} \phi_t(x)\, dx\, dt + \int_\Omega \nabla P_0(x) \phi_0(x)\, dx = 0. \tag{5.53}$$

- For a.e. $t \in (0, \infty)$ it holds

$$\int_\Omega \nabla \psi(x) \cdot u_t(x)\, dx = 0 \qquad \text{for all } \psi \in C^\infty_c(\Omega). \tag{5.54}$$

This definition is the classical notion of distributional solution for (5.4) except for the fact that the boundary condition $u_t \cdot \nu_\Omega = 0$ is not taken into account. In this sense it may look natural to consider $\psi \in C^\infty(\overline{\Omega})$ in (5.54), but since we are only able to prove that the velocity $u_t$ is locally in $L^1$, Equation (5.54) makes sense only with compactly supported $\psi$. On the other hand, exactly as in Section 5.2.3 we can build a measure preserving Lagrangian flow $F_t : \Omega \to \Omega$ associated to $u_t$, and such existence result can be interpreted as a very weak formulation of the constraint $u_t \cdot \nu_\Omega = 0$.

As pointed out to us by Cullen, this weak boundary condition is actually very natural: indeed, the classical boundary condition would prevent the formation of "frontal singularities" (which are physically expected to occur), *i.e.* the fluid initially at the boundary would not be able to move into the interior of the fluid, while this is allowed by our weak version of the boundary condition.

The following Theorem is the main result of this Section:

**Theorem 5.13.** *Let $\Omega \subseteq \mathbb{R}^3$ be a convex bounded open set, and let $\mathscr{L}^3_\Omega$ be the normalized Lebesgue measure restricted to $\Omega$. Let $\rho_0$ be a probability density on $\mathbb{R}^3$ such that $\rho_0 \in L^\infty(\mathbb{R}^3)$, $1/\rho_0 \in L^\infty_{\mathrm{loc}}(\mathbb{R}^3)$ and*

$$\limsup_{|x| \to \infty} \left( \rho_0(x) |x|^K \right) < \infty$$

*for some $K > 4$. Let $\rho_t$ be a solution of (5.5) given by Theorem 5.1, $P_t^* : \mathbb{R}^3 \to \mathbb{R}$ the unique convex function such that*

$$P_t^*(0) = 0 \quad \text{and} \quad (\nabla P_t^*)_\sharp(\rho_t \mathscr{L}^3) = \mathscr{L}^3_\Omega,$$

*and let $P_t : \mathbb{R}^3 \to \mathbb{R}$ be its convex conjugate.*

*Then the vector field $u_t$ in (5.9) is well defined, and the couple $(P_t, u_t)$ is a weak Eulerian solution of (5.4) in the sense of Definition 5.12.*

The proof of the above Theorem follows the same lines of the proof of Theorem 5.3. Indeed, once we have shown that the velocity field $u_t$ is in $L^1_{\mathrm{loc}}(\Omega)$ and the map $(x, t) \mapsto \nabla P_t^*(x)$ is in $W^{1,1}_{\mathrm{loc}}$, we can test (5.5) with the functions

$$\varphi_t(y) := y \phi_t(\nabla P_t^*(y)) \qquad \phi_t \in C_c^\infty(\Omega \times [0, \infty))$$

and

$$\varphi_t(y) := \phi(t) \psi(\nabla P_t^*(y)) \qquad \psi \in C_c^\infty(\Omega).$$

The same computations of the proof of Theorem 5.3 show that the couple $(P_t, u_t)$ is a distributional solution of (5.10). The rest of the Section is hence devoted to show the above mentioned regularity, summarized in Proposition 5.17.

We start recalling the following classical Theorem for optimal maps between convex sets. All the statements with the exception of the last one (which we comment below) can be deduced by the results in Chapter 2 and 3.

**Theorem 5.14. (Space regularity of optimal maps between convex sets).** *Let $\Omega_0$, $\Omega_1$ be open sets of $\mathbb{R}^3$, with $\Omega_1$ bounded and convex. Let $\mu = \rho \mathscr{L}^3$ and $\nu = \sigma \mathscr{L}^3$ be probability densities such that $\mu(\Omega_0) = 1$, $\nu(\Omega_1) = 1$. Assume that the density $\rho$ is locally bounded both from above and from below in $\Omega_0$, namely that for every compact set $K \subset \Omega$ there exist $\lambda_0 = \lambda_0(K)$ and $\Lambda_0 = \Lambda_0(K)$ satisfying*

$$0 < \lambda_0 \leq \rho(x) \leq \Lambda_0 \quad \forall x \in K.$$

*Futhermore, suppose that $\lambda_1 \leq \sigma(x) \leq \Lambda_1$ in $\Omega_1$. Then the following properties hold true.*

(i) *There exists a unique optimal transport map between $\mu$ and $\nu$, namely a unique (up to an additive constant) convex function $P^* : \Omega_0 \to \mathbb{R}$ such that $(\nabla P^*)_{\sharp} \mu = \nu$. Moreover $P^*$ is a strictly convex Aleksandrov solution of*

$$\det D^2 P^*(x) = f(x), \quad \text{with } f(x) = \frac{\rho(x)}{\sigma(\nabla P^*(x))}.$$

(ii) *$P^* \in W^{2,1}_{\text{loc}}(\Omega_0) \cap C^{1,\beta}_{\text{loc}}(\Omega_0)$. More precisely, if $\Omega \Subset \Omega_0$ is an open set and $0 < \lambda \leq \rho(x) \leq \Lambda < \infty$ in $\Omega$, then for any $k \in \mathbb{N}$ there exist constants $C_1 = C_1(k, \Omega, \Omega_1, \lambda, \Lambda, \lambda_1, \Lambda_1)$, $\beta = \beta(\lambda, \Lambda, \lambda_1, \Lambda_1)$, and $C_2 = C_2(\Omega, \Omega_1, \lambda, \Lambda, \lambda_1, \Lambda_1)$ such that*

$$\int_{\Omega} |D^2 P^*| \log^k_+ |D^2 P^*| \, dx \leq C_1,$$

*and*

$$\|P^*\|_{C^{1,\beta}(\Omega)} \leq C_2.$$

(iii) *Let us also assume that $\Omega_0$, $\Omega_1$ are bounded and uniformly convex, $\partial\Omega_0, \partial\Omega_1 \in C^{2,1}$, $\rho \in C^{1,1}(\Omega_0)$, $\sigma \in C^{1,1}(\Omega_1)$, and $\lambda_0 \leq \rho(x) \leq \Lambda_0$ in $\Omega_0$. Then*

$$P^* \in C^{3,\alpha}(\Omega_0) \cap C^{2,\alpha}(\overline{\Omega}_0) \quad \forall \alpha \in (0, 1),$$

106   Guido De Philippis

*and there exists a constant $C$ which depends only on $\alpha$, $\Omega_0$, $\Omega_1$, $\lambda_0$, $\lambda_1$, $\|\rho\|_{C^{1,1}}$, $\|\sigma\|_{C^{1,1}}$ such that*

$$\|P^*\|_{C^{3,\alpha}(\Omega_0)} \le C \quad \text{and} \quad \|P^*\|_{C^{2,\alpha}(\overline{\Omega_0})} \le C.$$

*Moreover, there exist positive constants $c_1$ and $c_2$ and $\kappa$, depending only on $\lambda_0$, $\lambda_1$, $\|\rho\|_{C^{0,\alpha}}$, and $\|\sigma\|_{C^{0,\alpha}}$, such that*

$$c_1 \operatorname{Id} \le D^2 P^*(x) \le c_2 \operatorname{Id} \quad \forall x \in \Omega_0$$

*and*

$$\nu_{\Omega_1}(\nabla P^*(x)) \cdot \nu_{\Omega_0}(x) \ge \kappa \quad \forall x \in \partial\Omega_0. \tag{5.55}$$

Some comments are in order. The regularity up to the boundary and the oblique derivative condition of the third statement have been proven by Caffarelli [24] and Urbas [90], in particular the oblique derivative condition (5.55) shall be thought as a "ellipticity" condition up to the boundary. More precisely, under this assumption the linearized operator associated to the *second boundary value* problem for the Monge-Ampère equation:

$$\begin{cases} \det D^2 u = f & \text{in } \Omega_1 \\ \nabla u(\Omega_1) = \Omega_2, \end{cases}$$

is strongly elliptic (see the proof of Lemma 5.15 below). This allows to use the continuity method outlined in Section 2.3 to find existence of smooth solutions of the above equation, see [90].

Finally we notice that in point (ii) we have only mentioned the $L \log L$ integrability of $D^2 P^*$ obtained in Theorem 3.1, clearly the optimal statement in view of Theorem 3.2 would have been:

(ii') *For every $\Omega \Subset \Omega_0$ and $0 < \lambda \le \rho(x) \le \Lambda < \infty$ in $\Omega$, there exist $\gamma(\lambda, \Lambda, \lambda_1, \Lambda_1) > 1$ and $C(\Omega, \Omega_1, \lambda, \Lambda, \lambda_1, \Lambda_1)$, such that*

$$\int_\Omega |D^2 u|^\gamma \le C.$$

However, due to the local bound $\log \rho \in L^\infty_{\text{loc}}$, this leads to an exponent $\gamma$ which depends on $\Omega$. and this makes the above estimates more complicated to use than (ii).

Here is the analogous of Lemma 5.7:

**Lemma 5.15 (Space-time regularity of transport).** *Let $\Omega \subseteq \mathbb{R}^3$ be a uniformly convex bounded domain with $\partial\Omega \in C^{2,1}$, let $R > 0$, and consider $\rho \in C^\infty(\overline{B(0,R)} \times [0,\infty))$ and $\mathcal{U} \in C_c^\infty(B(0,R) \times [0,\infty); \mathbb{R}^3)$ satisfying*

$$\partial_t \rho_t + \nabla \cdot (\mathcal{U}_t \rho_t) = 0 \quad \text{in } B(0,R) \times [0,\infty).$$

Assume that $\int_{B(0,R)} \rho_0 \, dx = 1$, and that for every $T > 0$ there exist $\lambda_T$ and $\Lambda_T$ such that

$$0 < \lambda_T \le \rho_t(x) \le \Lambda_T < \infty \qquad \forall (x,t) \in B(0,R) \times [0,T].$$

Consider the convex conjugate maps $P_t$ and $P_t^*$ such that $(\nabla P_t)_\sharp \mathscr{L}_\Omega^3 = \rho_t$ and $(\nabla P_t^*)_\sharp \rho_t = \mathscr{L}_\Omega^3$ (unique up to additive constants in $\Omega$ and $B(0,R)$ respectively). Then:

(i) $P_t^* - \fint_{B(0,R)} P_t^* \in \mathrm{Lip}_{\mathrm{loc}}([0,\infty); C^{2,\alpha}(\overline{B(0,R)}))$.
(ii) The following linearized Monge-Ampère equation holds for every $t \in [0,\infty)$:

$$\begin{cases} \nabla \cdot \left( \rho_t (D^2 P_t^*)^{-1} \partial_t \nabla P_t^* \right) = -\nabla \cdot (\rho_t \mathcal{U}_t) & \text{in } B(0,R) \\ \rho_t (D^2 P_t^*)^{-1} \partial_t \nabla P_t^* \cdot \nu = 0 & \text{on } \partial B(0,R). \end{cases} \qquad (5.56)$$

*Proof.* Observe that because $\rho_t$ solves a continuity equation with a smooth compactly supported vector field, $\int_{B(0,R)} \rho_t \, dx = 1$ for all $t$. Let us fix $T > 0$. From the regularity theory for the Monge-Ampére equation (Theorem 5.14 applied to $P_t$ and $P_t^*$) we obtain that $P_t \in C^{3,\alpha}(\Omega) \cap C^{2,\alpha}(\overline{\Omega})$ and $P_t^* \in C^{3,\alpha}(B(0,R)) \cap C^{2,\alpha}(\overline{B(0,R)})$ for every $\alpha \in (0,1)$, uniformly for $t \in [0,T]$, and there exist constants $c_1, c_2 > 0$ such that

$$c_1 \, \mathrm{Id} \le D^2 P_t^*(x) \le c_2 \, \mathrm{Id} \qquad \forall (x,t) \in B(0,R) \times [0,T]. \qquad (5.57)$$

Let $h \in C^{2,1}(\mathbb{R}^3)$ be a convex function such that $\Omega = \{y : h(y) < 0\}$ and $|\nabla h(y)| = 1$ on $\partial\Omega$, so that $\nabla h(y) = \nu_\Omega(y)$. Since $\nabla P_t^* \in C^{1,\alpha}(\overline{B(0,R)})$, it is a diffeomorphism onto its image, we have

$$h(\nabla P_t^*(x)) = 0 \qquad \forall (x,t) \in \partial B(0,R) \times [0,T]. \qquad (5.58)$$

As in Lemma 5.7, we investigate the time regularity of $P_t^* - \fint_{B(0,R)} P_t^*$. Possibly adding a time dependent constant to $P_t$, we can assume without loss of generality that $\int_{B(0,R)} P_t^* = 0$ for all $t$. By the condition $(\nabla P_t^*)_\sharp \rho_t = \mathscr{L}_\Omega^3$, arguing as in Lemma 5.7, we get that for any $0 \le s, t \le T$ and $x \in B(0,R)$ it holds

$$\frac{\rho_s - \rho_t}{s - t} = \sum_{i,j=1}^3 \left( \int_0^1 M_{ij}(\tau D^2 P_s^* + (1-\tau)D^2 P_t^*) \, d\tau \right) \partial_{ij} \left( \frac{P_s^* - P_t^*}{s - t} \right), \qquad (5.59)$$

where $M(D^2 P^2)$ is the cofactor matrix of $D^2 P^*$ which satisfies:

$$c_1^2 \, \mathrm{Id} \le \int_0^1 M_{ij}(\tau D^2 P_s^* + (1-\tau)D^2 P_t^*) \, d\tau \le c_2^2 \, \mathrm{Id}.$$

Moreover, from (5.58) we obtain that on $\partial B(0, R)$

$$
\begin{aligned}
0 &= \frac{h(\nabla P_s^*(x)) - h(\nabla P_t^*(x))}{s - t} \\
&= \int_0^1 \nabla h(\tau \nabla P_s^*(x) + (1 - \tau)\nabla P_t^*(x))\, d\tau \cdot \frac{\nabla P_s^*(x) - \nabla P_t^*(x)}{s - t}.
\end{aligned}
\tag{5.60}
$$

Also, from Theorem 5.14(iii) the oblique derivative condition holds, namely there exists $\kappa > 0$ such that

$$
\nabla h(\nabla P_t^*(x)) \cdot \nu_{B(0,R)}(x) \geq \kappa \qquad \forall x \in \partial B(0, R).
$$

Thus, since

$$
\lim_{s \to t} \int_0^1 \nabla h(\tau \nabla P_s^*(x) + (1 - \tau)\nabla P_t^*(x))\, d\tau = \nabla h(\nabla P_t^*(x))
$$

uniformly in $t$ and $x$, we have that

$$
\int_0^1 \nabla h(\tau \nabla P_s^*(x) + (1 - \tau)\nabla P_t^*(x))\, d\tau \cdot \nu_{B(0,R)}(x) \geq \frac{\kappa}{2}
$$

for $s - t$ small enough.

Hence, from the regularity theory for the oblique derivative problem [66, Theorem 6.30] we obtain that for any $\alpha \in (0, 1)$ there exists a constant $C$ depending only on $\Omega$, $T$, $\alpha$, $\|(\rho_s - \rho_t)/(s - t)\|_{C^{0,\alpha}(B(0,R))}$, such that

$$
\left\| \frac{P_s^*(x) - P_t^*(x)}{s - t} \right\|_{C^{2,\alpha}(\overline{B(0,R)})} \leq C.
$$

Since $\partial_t \rho_t \in L^\infty([0, T], C^{0,\alpha}(B(0, R)))$, this proves point (i) in the statement. To prove the second part, we let $s \to t$ in (5.59) to obtain, as in Lemma 5.7, Equation (5.56):

$$
-\nabla \cdot (\mathcal{U}_t \rho_t) = \nabla \cdot \left( \rho_t (D^2 P_t^*)^{-1} \partial_t \nabla P_t^* \right).
$$

In order to obtain the boundary condition in (5.56), we take to the limit as $s \to t$ in (5.60) to get

$$
\nabla h(\nabla P_t^*(x)) \cdot \partial_t \nabla P_t^*(x) = 0. \tag{5.61}
$$

Since $h$ satisfies $\Omega = \{y : h(y) < 0\}$ and $\nabla P_t^*$ maps $B(0, R)$ in $\Omega$, we have that $B(0, R) = \{y : h(\nabla P_t^*(y)) < 0\}$. Hence $\nu_{B(0,R)}(x)$ is proportional to $\nabla[h \circ \nabla P_t^*](x) = D^2 P_t^*(x)\nabla h(\nabla P_t^*(x))$, which implies that the exterior normal to $\Omega$ at point $\nabla P_t^*(x)$, which is $\nabla h(\nabla P_t^*(x))$, is collinear with $\rho_t(D^2 P_t^*)^{-1} v_{B(0,R)}$. Hence from (5.61) it follows that

$$\rho_t(D^2 P_t^*)^{-1} v_{B(0,R)} \cdot \partial_t \nabla P_t^* = 0,$$

as desired.  $\square$

**Lemma 5.16 (Decay estimates on $\rho_t$).** *Let $v_t : \mathbb{R}^3 \times [0, \infty) \to \mathbb{R}^3$ be a $C^\infty$ velocity field and suppose that*

$$\sup_{x,t} |\nabla \cdot v_t(x)| \leq N, \qquad |v_t(x)| \leq A|x| + D \quad \forall (x, t) \in \mathbb{R}^3 \times [0, \infty)$$

*for suitable constants $N$, $A$, $D$. Let $\rho_0$ be a probability density, and let $\rho_t$ be the solution of the continuity equation*

$$\partial_t \rho_t + \nabla \cdot (v_t \rho_t) = 0 \qquad in \ \mathbb{R}^3 \times (0, \infty) \tag{5.62}$$

*starting from $\rho_0$. Then:*

(i) *For every $r > 0$ and $t \in [0, \infty)$ it holds*

$$\|\rho_t\|_\infty \leq e^{Nt} \|\rho_0\|_\infty, \tag{5.63}$$

$$\rho_t(x) \geq e^{-Nt} \inf\left\{ \rho_0(y) : y \in B\left(0, re^{At} + D\frac{e^{At}-1}{A}\right) \right\} \quad \forall x \in B(0,r). \tag{5.64}$$

(ii) *Let us assume that there exist $d_0 \in [0, \infty)$ and $M \in [0, \infty)$ such that*

$$\rho_0(x) \leq \frac{d_0}{|x|^K} \qquad whenever \quad |x| \geq M. \tag{5.65}$$

*Then for every $t \in [0, \infty)$ we have that*

$$\rho_t(x) \leq \frac{d_0 2^K e^{(N+AK)t}}{|x|^K} \ whenever \ |x| \geq 2Me^{At} + 2D\frac{e^{At}-1}{A}. \tag{5.66}$$

(iii) *Let us assume that there exists $R > 0$ such that $\rho_0$ is smooth in $\overline{B(0, R)}$, vanishes outside $\overline{B(0, R)}$, and that $v_t$ is compactly supported inside $B(0, R)$ for all $t \geq 0$. Then $\rho_t$ is smooth inside $\overline{B(0, R)}$ and vanishes outside $\overline{B(0, R)}$ for all $t \geq 0$. Moreover if $0 < \lambda \leq \rho_0 \leq \Lambda < \infty$ inside $B(0, R)$, then*

$$\lambda e^{-tN} \leq \rho_t \leq \Lambda e^{tN} \quad inside \ B(0, R) \ for \ all \ t \geq 0. \tag{5.67}$$

*Proof.* Let $X_t(x) \in C^\infty(\mathbb{R}^3 \times [0, \infty))$ be the flow associated to the velocity field $\boldsymbol{v}_t$, namely the solution to

$$\begin{cases} \dfrac{d}{dt} X_t(x) = \boldsymbol{v}_t(X_t(x)) \\ X_0(x) = x. \end{cases} \tag{5.68}$$

For every $t \geq 0$ the map $t \mapsto X_t(x)$ is invertible in $\mathbb{R}^3$, with inverse denoted by $X_t^{-1}$.

The solution to the continuity equation (5.62) is given by $\rho_t = X_{t\sharp}\rho_0$, and from the well-known theory of characteristics it can be written explicitly using the flow:

$$\rho_t(x) = \rho_0(X_t^{-1}(x)) e^{\int_0^t \nabla \cdot v_s(X_s(X_t^{-1}(x)))\, ds} \qquad \forall (x, t) \in \mathbb{R}^3 \times [0, \infty). \tag{5.69}$$

Since the divergence is bounded, we therefore obtain

$$\rho_0(X_t^{-1}(x)) e^{-Nt} \leq \rho_t(x) \leq \rho_0(X_t^{-1}(x)) e^{Nt} \tag{5.70}$$

Now we deduce the statements of the lemma from the properties of the flow $X_t$.

(i) From (5.70) we have that

$$\rho_t(x) \leq e^{Nt} \rho_0(X_t^{-1}(x)) \leq e^{Nt} \sup_{x \in \mathbb{R}^3} \rho_0(x),$$

which proves (5.63). From the equation (5.68) we obtain

$$\left| \frac{d}{dt} |X_t(x)| \right| \leq |\partial_t X_t(x)| \leq A|X_t(x)| + D$$

which can be rewritten as

$$-A|X_t(x)| - D \leq \frac{d}{dt} |X_t(x)| \leq A|X_t(x)| + D. \tag{5.71}$$

From the first inequality we get

$$|X_t(x)| \geq |x| e^{-At} - D \frac{1 - e^{-At}}{A},$$

which implies

$$|x| e^{At} + D \frac{e^{At} - 1}{A} \geq |X_t^{-1}(x)|,$$

or equivalently

$$X_t^{-1}(\{|x| \leq r\}) \subseteq \left\{ |x| \leq re^{At} + D \frac{e^{At} - 1}{A} \right\}. \tag{5.72}$$

Hence from (5.70) and (5.72) we obtain that, for every $x \in B(0, r)$,

$$\begin{aligned} \rho_t(x) &\geq e^{-Nt} \rho_0(X_t^{-1}(x)) \\ &\geq e^{-Nt} \inf\{\rho_0(y) : y \in X_t^{-1}(B_r(0))\} \\ &\geq e^{-Nt} \inf\left\{\rho_0(y) : |y| \leq re^{At} + D\frac{e^{At} - 1}{A}\right\}, \end{aligned}$$

which proves (5.64).

(ii) From the second inequality in (5.71), we infer

$$|X_t(x)| \leq |x|e^{At} + D\frac{e^{At} - 1}{A},$$

which implies

$$|x| \leq |X_t^{-1}(x)|e^{At} + D\frac{e^{At} - 1}{A}. \tag{5.73}$$

Thus, if $|x| \geq 2Me^{At} + 2D\frac{e^{At}-1}{A}$, we easily deduce from (5.73) that $|X_t^{-1}(x)| \geq M + |x|e^{-At}/2$, so by (5.65)

$$\rho_t(x) \leq e^{Nt} \rho_0(X_t^{-1}(x)) \leq \frac{d_0 e^{Nt}}{|X_t^{-1}(x)|^K} \leq \frac{d_0 2^K e^{(N+AK)t}}{|x|^K},$$

which proves (5.66).

(iii) If $v_t = 0$ in a neighborhood of $\partial B(0, R)$ it can be easily verified that the flow maps $X_t : \mathbb{R}^3 \to \mathbb{R}^3$ leave both $B(0, R)$ and its complement invariant. Moreover the smoothness of $v_t$ implies that also $X_t$ is smooth. Therefore all the properties of $\rho_t$ follow directly from (5.69). $\qquad \square$

We are now ready to prove the regularity of $\nabla P_t^*$, as we explained the proof of Theorem 5.13 easily follows from this.

**Proposition 5.17 (Time regularity of optimal maps).** *Let $\Omega \subseteq \mathbb{R}^3$ be a bounded, convex, open set and let $\mathrm{diam}(\Omega)$ be such that $\overline{\Omega} \subset B(0, \mathrm{diam}(\Omega))$. Let $\rho_t$ and $P_t$ be as in Theorem 5.1, in addition let us assume that there exist $K > 4$, $M \geq 0$ and $c_0 > 0$ such that*

$$\rho_0(x) \leq \frac{c_0}{|x|^K} \qquad \text{whenever } |x| \geq M. \tag{5.74}$$

*Then $\nabla P_t^* \in W_{loc}^{1,1}(\mathbb{R}^3 \times [0, \infty); \mathbb{R}^3)$. Moreover for every $k \in \mathbb{N}$ and $T > 0$ there exists a constant $C = C(k, T, M, c_0, \|\rho_0\|_\infty, \mathrm{diam}(\Omega))$ such that, for almost every $t \in [0, T]$ and for all $r \geq 0, t$ it holds*

$$\int_{B(0,r)} \rho_t |\partial_t \nabla P_t^*| \log_+^k (|\partial_t \nabla P_t^*|) \, dx$$

$$\leq 2^{3(k-1)} \int_{B(0,r)} \rho_t |D^2 P_t^*| \log_+^{2k} (|D^2 P_t^*|) \, dx + C \tag{5.75}$$

*Proof. Step* 1 : *The smooth case.* In the first part of the proof we assume that $\Omega$ is a convex smooth domain, and, besides (5.74), that for some $R > 0$ the following additional properties hold:

$$\rho_t \in C^\infty(\overline{B(0, R)} \times \mathbb{R}), \quad \mathcal{U}_t \in C_c^\infty(B(0, R) \times \mathbb{R}; \mathbb{R}^3), \quad (5.76)$$

$$|\nabla \cdot \mathcal{U}_t| \leq N$$

$$\lambda 1_{B(0,R)}(x) \leq \rho_0(x) \leq \Lambda 1_{B(0,R)}(x) \quad \forall x \in \mathbb{R}^3, \quad (5.77)$$

$$\partial_t \rho_t + \nabla \cdot (\mathcal{U}_t \rho_t) = 0 \quad \text{in } \mathbb{R}^3 \times [0, \infty), \quad (5.78)$$

$$(\nabla P_t^*)_\sharp \rho_t = \mathscr{L}_\Omega^3, \quad (5.79)$$

$$|\mathcal{U}_t(x)| \leq |x| + \text{diam}(\Omega) \quad (5.80)$$

for some constants $N$, $\lambda$, $\Lambda$, and we prove that (5.15) holds for every $t \in [0, T]$. Notice that in this step we do not assume any coupling between the velocity $\mathcal{U}_t$ and the transport map $\nabla P_t^*$. In the second step we prove the general case through an approximation argument.

Let us assume that the regularity assumptions (5.77) through (5.80) hold. By Lemma 5.16 we infer that, for any $T > 0$, there exist positive constants $\lambda_T, \Lambda_T, c_T, M_T$, with $M_T \geq 1$, such that

$$\lambda_T 1_{B(0,R)}(x) \leq \rho_t(x) \leq \Lambda_T 1_{B(0,R)}(x), \quad (5.81)$$

$$\rho_t(x) \leq \frac{c_T}{|x|^K} \quad \text{for } |x| \geq M_T, \quad \text{for all } t \in [0, T]. \quad (5.82)$$

By Lemma 5.15 we have that $\partial_t P_t^* \in C^2(\overline{B(0, R)})$, and it solves

$$\begin{cases} \nabla \cdot \left(\rho_t (D^2 P_t^*)^{-1} \partial_t \nabla P_t^*\right) = -\nabla \cdot (\rho_t \mathcal{U}_t) & \text{in } B(0, R) \\ \rho_t (D^2 P_t^*)^{-1} \partial_t \nabla P_t^* \cdot \nu = 0 & \text{in } \partial B(0, R). \end{cases} \quad (5.83)$$

Multiplying (5.83) by $\partial_t P_t^*$ and integrating by parts, we get

$$\int_{B(0,R)} \rho_t |(D^2 P_t^*)^{-1/2} \partial_t \nabla P_t^*|^2 \, dx = \int_{B(0,R)} \rho_t \partial_t \nabla P_t^* \cdot (D^2 P_t^*)^{-1} \partial_t \nabla P_t^* \, dx$$

$$= -\int_{B(0,R)} \rho_t \partial_t \nabla P_t^* \cdot \mathcal{U}_t \, dx. \quad (5.84)$$

(Notice that, thanks to the boundary condition in (5.83), we do not have any boundary term in (5.84).) From Cauchy-Schwartz inequality

$$-\int_{B(0,R)} \rho_t \partial_t \nabla P_t^* \cdot (D^2 P_t^*)^{-1/2} (D^2 P_t^*)^{1/2} \mathcal{U}_t \, dx$$

$$\leq \left(\int_{B(0,R)} \rho_t |(D^2 P_t^*)^{-1/2} \partial_t \nabla P_t^*|^2 dx\right)^{1/2} \left(\int_{B(0,R)} \rho_t |(D^2 P_t^*)^{1/2} \mathcal{U}_t|^2 dx\right)^{1/2}. \quad (5.85)$$

Moreover, the second term in the right-hand side of (5.85) is controlled by

$$\int_{B(0,R)} \rho_t \mathcal{U}_t \cdot D^2 P_t^* \mathcal{U}_t \, dx \leq \max_{B(0,R)} \left( \rho_t^{1/2} |\mathcal{U}_t|^2 \right) \int_{B(0,R)} \rho_t^{1/2} |D^2 P_t^*| \, dx.$$
(5.86)

Hence we obtain, as in Proposition 5.6,

$$\int_{B(0,R)} \rho_t |(D^2 P_t^*)^{-1/2} \partial_t \nabla P_t^*|^2 \, dx$$
$$\leq \max_{B(0,R)} \left( \rho_t^{1/2} |\mathcal{U}_t|^2 \right) \int_{B(0,R)} \rho_t^{1/2} |D^2 P_t^*| \, dx.$$
(5.87)

Notice however that we will need to send $R$ to infinity at the end of the proof, for this reason we need a bound on the above quantity independent on $R$.

From (5.80), (5.81), and (5.82) we estimate the first factor as follows:

$$\max_{|x| \leq M_T} \left( \rho_t^{1/2}(x) |\mathcal{U}_t(x)|^2 \right) \leq \Lambda_T^{1/2} (M_T + \text{diam}(\Omega))^2,$$
(5.88)

$$\max_{M_T \leq |x|} \left( \rho_t^{1/2}(x) |\mathcal{U}_t(x)|^2 \right) \leq \max_{M_T \leq |x|} \left\{ \frac{\sqrt{c_T}}{|x|^{K/2}} (|x| + \text{diam}(\Omega))^2 \right\},$$
(5.89)

and the latter term is finite because $M_T \geq 1$ and $K > 4$.

In order to estimate the second factor, we observe that since $D^2 P_t^*$ is a nonnegative matrix the estimate $|D^2 P_t^*| \leq 3 \Delta P_t^*$ holds. Hence, by (5.81) and (5.82) we obtain

$$\int_{B(0,R)} \rho_t^{1/2} |D^2 P_t^*| \, dx \leq \int_{\{|x| \leq M_T\}} \rho_t^{1/2} |D^2 P_t^*| \, dx$$
$$+ \int_{\{|x| > M_T\}} \rho_t^{1/2} |D^2 P_t^*| \, dx$$
$$\leq 3 \int_{\{|x| \leq M_T\}} \Lambda_T^{1/2} \Delta P_t^* \, dx$$
$$+ 3 \int_{\{|x| > M_T\}} \frac{\sqrt{c_T}}{|x|^{K/2}} \Delta P_t^* \, dx.$$

The second integral can be rewritten as

$$\int_0^\infty \int_{\{|x| > M_T\} \cap \{|x|^{-K/2} > s\}} \Delta P_t^* \, dx \, ds,$$

which is bounded by

$$\int_0^{[M_T]^{-K/2}} ds \int_{\{|x|\le s^{-2/K}\}} \Delta P_t^* \, dx.$$

From the divergence formula, since $|\nabla P_t^*(x)| \le \text{diam}(\Omega)$ (because $\nabla P_t^*(x) \in \Omega$ for every $x \in \mathbb{R}^3$) and $M_T \ge 1$ (so $[M_T]^{-K/2} \le 1$) we obtain

$$\int_{B(0,R)} \rho_t^{1/2} |D^2 P_t^*| \, dx \le 3\Lambda_T^{1/2} \int_{\{|x|=M_T\}} |\nabla P_t^*| \, d\mathcal{H}^2$$

$$+ 3\sqrt{c_T} \int_0^{[M_T]^{-K/2}} ds \int_{\{|x|=s^{-2/K}\}} |\nabla P_t^*| \, d\mathcal{H}^2$$

$$\le 12\pi \Lambda_T^{1/2} M_T^2 \text{diam}(\Omega)$$

$$+ 12\pi \sqrt{c_T} \text{diam}(\Omega) \int_0^1 s^{-4/K} \, ds \tag{5.90}$$

for all $t \in [0, T]$. Since $K > 4$ the last integral is finite, so the right-hand side is bounded and we obtain a global-in-space estimate on the left-hand side.

Thus, from (5.87), (5.88), (5.89), and (5.90), it follows that there exists a constant $C_1 = C_1(T, M, c_0, \Lambda, \text{diam}(\Omega))$ (notice that the constant does not depend on the lower bound on the density) such that

$$\int_{B(0,R)} \rho_t |(D^2 P_t^*)^{-1/2} \partial_t \nabla P_t^*|^2 \, dx \le C_1. \tag{5.91}$$

Applying now the inequality (see Lemma 5.18 below)

$$ab \log_+^k(ab) \le 2^{k-1}\left[\left(\frac{k}{e}\right)^k + 1\right] b^2 + 2^{3(k-1)} a^2 \log_+^{2k}(a) \qquad \forall\, a, b \ge 0,$$

with $a = |(D^2 P_t^*)^{1/2}|$ and $b = |(D^2 P_t^*)^{-1/2}\partial_t \nabla P_t^*(x)|$ we deduce the existence of a constant $C_2 = C_2(k)$ such that

$$|\partial_t \nabla P_t^*| \log_+^k(|\partial_t \nabla P_t^*|) \le 2^{3(k-1)}|(D^2 P_t^*)^{1/2}|^2 \log_+^{2k}(|(D^2 P_t^*)^{1/2}|^2)$$

$$+ C_2 |(D^2 P_t^*)^{-1/2}\partial_t \nabla P_t^*|^2$$

$$= 2^{3(k-1)}|D^2 P_t^*| \log_+^{2k}(|D^2 P_t^*|)$$

$$+ C_2 |(D^2 P_t^*)^{-1/2}\partial_t \nabla P_t^*|^2.$$

Integrating the above inequality over $B(0, r)$ and using (5.91), we finally obtain

$$
\begin{aligned}
&\int_{B(0,r)} \rho_t |\partial_t \nabla P_t^*| \log_+^k (|\partial_t \nabla P_t^*|) \, dx \\
&\leq 2^{3(k-1)} \int_{B(0,r)} \rho_t |D^2 P_t^*| \log_+^{2k} (|D^2 P_t^*|) \, dx \\
&\quad + C_2 \int_{B(0,R)} \rho_t |(D^2 P_t^*)^{-1/2} \partial_t \nabla P_t^*|^2 \, dx \\
&\leq 2^{3(k-1)} \int_{B(0,r)} \rho_t |D^2 P_t^*| \log_+^{2k} (|D^2 P_t^*|) \, dx + C_1 \cdot C_2,
\end{aligned}
\tag{5.92}
$$

for all $0 < r \leq R$.

*Step* 2 : *The approximation argument.* We now consider the velocity field $\mathcal{U}$ given by Theorem 5.1, we take a sequence of smooth convex domains $\Omega_n$ which converges to $\Omega$ in the Hausdorff distance, and a sequence $(\psi^n) \subset C_c^\infty(B(0, n))$ of cut off functions such that $0 \leq \psi_n \leq 1$, $\psi^n(x) = 1$ inside $B(0, n/2)$, $|\nabla \psi^n| \leq 2/n$ in $\mathbb{R}^3$. Let us also consider a sequence of space-time mollifiers $\sigma^n$ with support contained in $B(0, 1/n)$ and a sequence of space mollifiers $\varphi^n$. We extend the function $\mathcal{U}_t$ for $t \leq 0$ by setting $\mathcal{U}_t = 0$ for every $t < 0$.

Let us consider a compactly supported space regularization of $\rho_0$ and a space-time regularization of $U$, namely

$$
\rho_0^n := \frac{(\rho_0 * \varphi^n)}{c_n} 1_{B(0,n)}, \qquad \mathcal{U}_t^n(x) := (\mathcal{U} * \sigma^n) \psi^n,
$$

where $c_n \uparrow 1$ is chosen so that $\rho_0^n$ is a probability measure on $\mathbb{R}^3$. Let $\rho_t^n$ be the solution of the continuity equation

$$
\partial_t \rho_t^n + \nabla \cdot (\mathcal{U}_t^n \rho_t^n) = 0 \qquad \text{in } \mathbb{R}^3 \times [0, \infty)
$$

with initial datum $\rho_0^n$. From the regularity of the velocity field $\mathcal{U}_t^n$ and of the initial datum $\rho_0^n$ we have that $\rho^n \in C^\infty(B(0, n) \times [0, \infty))$.

Since $\mathcal{U}_t$ is divergence-free and satisfies the inequality $|\mathcal{U}_t(x)| \leq |x| +$ diam$(\Omega)$, we get

$$
|\mathcal{U}_t^n|(x) \leq |\mathcal{U} * \sigma^n|(x) \leq \|\mathcal{U}_t\|_{L^\infty(B(x,1/n))} \leq |x| + \text{diam}(\Omega)
$$

$$
+ \frac{1}{n} \leq |x| + \text{diam}(\Omega) + 1,
$$

$$
|\nabla \cdot \mathcal{U}_t^n|(x) = |(\mathcal{U}_t * \sigma^n) \cdot \nabla \psi^n|(x) \leq \frac{2(n + 1 + \text{diam}(\Omega))}{n} \leq 3
$$

for $n$ large enough. Moreover, from the properties of $\rho_0$ we obtain that, for $n$ large enough,

$$\rho_0^n(x) \leq \frac{2c_0}{(|x| - 1/n)^K} \leq \frac{4c_0}{|x|^K} \qquad \forall\, |x| \geq M + 2,$$

$$\|\rho_0^n\|_\infty \leq 2\|\rho_0\|_\infty \quad \text{and} \quad \left\|\frac{1}{\rho_0^n}\right\|_{L^\infty(B(0,n))} \leq \left\|\frac{1}{\rho_0}\right\|_{L^\infty(B(0,n+1))}.$$

Hence the hypotheses of Lemma 5.16 are satisfied with $N = 3$, $A = 1$, $D = \operatorname{diam}(\Omega) + 1$, $d_0 = 4c_0$. Moreover $\rho_t^n$ vanishes outside $B(0, n)$, and by (5.67) there exist constants $\lambda_n := e^{-3T}\|1/\rho_0\|_{L^\infty(B(0,n+1))}^{-1} > 0$, $\Lambda := 2e^{3T}\|\rho_0\|_\infty$, and $M_1$, $c_1$ depending on $T, M, c_0, \operatorname{diam}(\Omega)$ only, such that

$$\lambda_n \leq \rho_t^n(x) \leq \Lambda \qquad \forall\, (x, t) \in B(0, n) \times [0, T],$$

$$\rho_t^n(x) \leq \frac{c_1}{|x|^K} \qquad \text{whenever } |x| \geq M_1.$$

(Observe that $\lambda_n$ depends on $n$, but the other constants are all independent of $n$.) Thus, from Statement (ii) of Lemma 5.16 we get that, for all $r > 0$,

$$\rho_t^n(x) \geq e^{-3T} \inf\left\{\rho_0^n(y) : y \in B\left(0, re^t + (\operatorname{diam}(\Omega)+1)[e^t - 1]\right)\right\}$$
$$\forall\, (x, t) \in B(0, r) \times [0, T]. \quad (5.93)$$

If $n$ is large enough, the right-hand side of (5.93) is different from 0, and it can be estimated from below in terms of $\rho_0$ by

$$\lambda = \lambda(r, T, \rho_0, \Omega)$$
$$:= e^{-3T} \inf\left\{\rho_0(y) : y \in B\left(0, re^t + (\operatorname{diam}(\Omega)+1)[e^t - 1] + 1\right)\right\} > 0.$$

Therefore, for any $r > 0$ we can bound the density $\rho^n$ from below inside $B(0, r)$ with a constant *independent of n*:

$$\lambda \leq \rho_t^n(x) \leq \Lambda \qquad \forall\, (x, t) \in B(0, r) \times [0, T]. \quad (5.94)$$

Let now $P_t^{n*}$ be the unique convex function such that $P_t^{n*}(0) = 0$ and $(\nabla P_t^n)_\sharp \rho_t^n = \mathscr{L}_{\Omega_n}^3$. From the stability of solutions to the continuity equation with $BV$ velocity field, [2, Theorem 6.6], we infer that

$$\rho_t^n \to \rho_t \qquad \text{in } L^1_{\text{loc}}(\mathbb{R}^3), \text{ for any } t > 0, \quad (5.95)$$

where $\rho_t$ is the unique solution of (5.5) corresponding to the velocity field $\mathcal{U}$. Moreover by the stability of optimal maps

$$\nabla P_t^{n*} \to \nabla P_t^* \qquad \text{in } L^1_{\text{loc}}(\mathbb{R}^3) \quad (5.96)$$

for any $t > 0$. By Theorem 5.14 (ii) [9] and (5.94), for every $k \in \mathbb{N}$

$$\limsup_{n \to \infty} \int_{B(0,r)} \rho_t^n |D^2 P_t^{n*}| \log_+^{2k}(|D^2 P_t^{n*}|)\, dx < \infty \qquad \forall\, r > 0,$$

and by Theorem 4.1

$$\lim_{n \to \infty} \int_{B(0,r)} \rho_t^n |D^2 P_t^{n*}| \log_+^{2k}(|D^2 P_t^{n*}|)\, dx$$
$$= \int_{B(0,r)} \rho_t |D^2 P_t^*| \log_+^{2k}(|D^2 P_t^*|)\, dx \qquad \forall\, r > 0. \tag{5.97}$$

Since $(\rho_t^n, \mathcal{U}_t^n)$ satisfy the assumptions (5.77) through (5.80), by Step 1 we can apply (5.92) to $(\rho_t^n, \mathcal{U}_t^n)$ to obtain

$$\int_{B(0,r)} \rho_t^n |\partial_t \nabla P_t^{n*}| \log_+^{k}(|\partial_t \nabla P_t^{n*}|)\, dx$$
$$\leq 2^{3(k-1)} \int_{B(0,r)} \rho_t |D^2 P_t^{n*}| \log_+^{2k}(|D^2 P_t^{n*}|)\, dx + C \tag{5.98}$$

for all $r < n$, where the constant $C$ does not depend on $n$. At this point we obtain (5.75) from (5.98) and (5.97) arguing as in the last part of Proposition 5.6.   $\square$

**Lemma 5.18.** *For every* $k \in \mathbb{N}$ *we have*

$$ab \log_+^{k}(ab) \leq 2^{k-1} \left[ \left( \frac{k}{e} \right)^k + 1 \right] b^2 + 2^{3(k-1)} a^2 \log_+^{2k}(a) \qquad \forall\, a, b > 0. \tag{5.99}$$

*Proof.* From the elementary inequalities

$$\log_+(ts) \leq \log_+(t) + \log_+(s), \quad (t+s)^k \leq 2^{k-1}(t^k + s^k), \quad \log_+^{k}(t) \leq \left( \frac{k}{e} \right)^k t$$

---

[9] Recall that by Theorem 2.2 the modulus of strict convexity of the $(P^*)^n$ depends only on the limiting domain $\Omega$, in particular all the constants in the $L \log L$ estimates remain bounded

which hold for every $t$, $s > 0$, we infer

$$
\begin{aligned}
ab \log_+^k (ab) &\leq ab \left[ \log_+ \left( \frac{b}{a} \right) + 2 \log_+(a) \right]^k \\
&\leq 2^{k-1} ab \left[ \log_+^k \left( \frac{b}{a} \right) + 2^k \log_+^k(a) \right] \\
&\leq 2^{k-1} \left[ \left( \frac{k}{e} \right)^k b^2 + 2^k ab \log_+^k(a) \right] \\
&\leq 2^{k-1} \left[ \left( \frac{k}{e} \right)^k b^2 + b^2 + 2^{2(k-1)} a^2 \log_+^{2k}(a) \right],
\end{aligned}
$$

which proves (5.99). $\qquad\qquad\square$

# Chapter 6
# Partial regularity of optimal transport maps

The goal of this chapter (based on a joint work with Alessio Figalli [43]) is to prove partial regularity of optimal transport maps under mild assumptions on the cost function $c$ and on the densities $f$ and $g$, Theorems 6.1 and 6.2 below.

We recall here our assumptions on $c$ (see Section 1.3 for a discussion about existence and the main notation):

**(C0)** The cost function $c : X \times Y \to \mathbb{R}$ is of class $C^2$ with $\|c\|_{C^2(X \times Y)} < \infty$.
**(C1)** For any $x \in X$, the map $Y \ni y \mapsto -D_x c(x, y) \in \mathbb{R}^n$ is injective.
**(C2)** For any $y \in Y$, the map $X \ni x \mapsto -D_y c(x, y) \in \mathbb{R}^n$ is injective.
**(C3)** $\det(D_{xy} c)(x, y) \neq 0$ for all $(x, y) \in X \times Y$.

Our results can be stated as follows (cp. Theorem 1.30)

**Theorem 6.1.** *Let $X, Y \subset \mathbb{R}^n$ be two bounded open sets, $f : X \to \mathbb{R}^+$ and $g : Y \to \mathbb{R}^+$ be continuous probability densities bounded away from zero and infinity on $X$ and $Y$ respectively. Assume that the cost $c : X \times Y \to \mathbb{R}$ satisfies (C0)-(C3), and denote by $T$ the unique optimal transport map sending $f$ onto $g$. Then there exist two relatively closed sets $\Sigma_X \subset X$, $\Sigma_Y \subset Y$ of measure zero such that $T : X \setminus \Sigma_X \to Y \setminus \Sigma_Y$ is a homeomorphism of class $C_{loc}^{0,\beta}$ for any $\beta < 1$. In addition, if $c \in C_{loc}^{k+2,\alpha}(X \times Y)$, $f \in C_{loc}^{k,\alpha}(X)$, and $g \in C_{loc}^{k,\alpha}(Y)$ for some $k \geq 0$ and $\alpha \in (0, 1)$, then $T : X \setminus \Sigma_X \to Y \setminus \Sigma_Y$ is a diffeomorphism of class $C_{loc}^{k+1,\alpha}$.*

**Theorem 6.2.** *Let $M$ be a smooth Riemannian manifold, and let $f, g : M \to \mathbb{R}^+$ be two continuous probability densities, locally bounded away from zero and infinity on $M$. Let $T : M \to M$ denote the optimal transport map for the cost $c = d^2/2$ sending $f$ onto $g$. Then there exist two closed sets $\Sigma_X, \Sigma_Y \subset M$ of measure zero such that $T : M \setminus \Sigma_X \to M \setminus \Sigma_Y$ is a homeomorphism of class $C_{loc}^{0,\beta}$ for any $\beta < 1$. In addition, if both $f$ and $g$ are of class $C^{k,\alpha}$, then $T : M \setminus \Sigma_X \to M \setminus \Sigma_Y$ is a diffeomorphism of class $C_{loc}^{k+1,\alpha}$.*

The Chapter is structured as follows, in Section 6.1, we show how both Theorem 6.1 and Theorem 6.2 are a direct consequence of a local regularity results around differentiability points of $T$, see Theorems 6.5 and 6.11. Finally, Sections 6.2 and 6.3 are devoted to the proof of this local result.

## 6.1. The localization argument and proof of the results

The goal of this section is to prove Theorems 6.1 and 6.2 by showing that the assumptions of Theorems 6.5 and 6.11 below are satisfied near almost every point.

The rough idea is the following: if $\bar{x}$ is a point where the semiconvex function $u$ is twice differentiable, then around that point $u$ looks like a parabola. In addition, by looking close enough to $\bar{x}$, the cost function $c$ will be very close to the linear one and the densities will be almost constant there. Hence we can apply Theorem 6.5 to deduce that $u$ is of class $C^{1,\beta}$ in neighborhood of $\bar{x}$ (resp. $u$ is of class $C^{k+2,\alpha}$ by Theorem 6.11, if $c \in C_{\text{loc}}^{k+2,\alpha}$ and $f, g \in C_{\text{loc}}^{k,\alpha}$), which implies in particular that $T_u$ is of class $C^{0,\beta}$ in neighborhood of $\bar{x}$ (resp. $T_u$ is of class $C^{k+1,\alpha}$ by Theorem 6.11, if $c \in C_{\text{loc}}^{k+2,\alpha}$ and $f, g \in C_{\text{loc}}^{k,\alpha}$). Being our assumptions completely symmetric in $x$ and $y$, we can apply the same argument to the optimal map $T^*$ sending $g$ onto $f$. Since $T^* = (T_u)^{-1}$ (see the discussion below), it follows that $T_u$ is a global homeomorphism of class $C_{\text{loc}}^{0,\beta}$ (resp. $T_u$ is a global diffeomorphism of class $C_{\text{loc}}^{k+1,\alpha}$) outside a closed set of measure zero.

We now give a detailed proof.

*Proof of Theorem* 6.1. As discussed in Section 1.3, if we introduce the "$c$-conjugate" of $u$, that is the function $u^c : Y \to \mathbb{R}$ defined as

$$u^c(y) := \sup_{x \in X} \big\{ - c(x, y) - u(x) \big\}.$$

Then $u^c$ is $c^*$-convex, where

$$c^*(y, x) := c(x, y) \quad \text{and} \quad x \in \partial_{c^*} u^c(y) \quad \Leftrightarrow \quad y \in \partial_c u(x). \quad (6.1)$$

Being the assumptions on $c$ completely symmetric in $x$ and $y$, clearly also $c^*$ satisfies the same assumption as $c$. By Theorem 1.28, the optimal map $T^*$ (with respect to $c^*$) sending $g$ onto $f$ [1], is equal to

$$T_{u^c}(y) = \text{c*-exp}_y\big(\nabla u^c(y)\big).$$

---

[1] In the sequel we will identify an absolutely continuous measure with its density.

or equivalently

$$D_y c\big(T_{u^c}(y), y\big) = -\nabla u^c(y). \tag{6.2}$$

In addition (recall that $f$ and $g$ are bounded and strictly positive on their support), Theorem 1.28 asserts that

$$T_{u^c}\big(T_u(x)\big) = x, \quad T_u\big(T_{u^c}(y)\big) = y \quad \text{for a.e. } x \in X, y \in Y. \tag{6.3}$$

Since semiconvex functions are twice differentiable a.e., there exist sets $X_1 \subset X, Y_1 \subset Y$ of full measure such that (6.3) holds for every $x \in X_1$ and $y \in Y_1$, and in addition $u$ is twice differentiable for every $x \in X_1$ and $u^c$ is twice differentiable for every $y \in Y_1$. Let us define

$$X' := X_1 \cap (T_u)^{-1}(Y_1).$$

Using that $T_u$ transports $f$ on $g$ and that the two densities are bounded away from zero and infinity, we see that $X'$ is of full measure in $X$.

We fix a point $\bar{x} \in X'$. Since $u$ is differentiable at $\bar{x}$ (being twice differentiable), it follows by (1.33) that the set $\partial_c u(\bar{x})$ is a singleton, namely $\partial_c u(\bar{x}) = \{\text{c-exp}_{\bar{x}}(\nabla u(\bar{x}))\}$. Set $\bar{y} := \text{c-exp}_{\bar{x}}(\nabla u(\bar{x}))$. Since $\bar{y} \in Y_1$ (by definition of $X'$), $u^c$ is twice differentiable at $\bar{y}$ and $\bar{x} = T_{u^c}(\bar{y})$. Up to a translation in the system of coordinates (both in $x$ and $y$) we can assume that both $\bar{x}$ and $\bar{y}$ coincide with the origin $\mathbf{0}$.

Let us define

$$\bar{u}(z) := u(z) - u(\mathbf{0}) + c(z, \mathbf{0}) - c(\mathbf{0}, \mathbf{0}),$$
$$\bar{c}(z, w) := c(z, w) - c(z, \mathbf{0}) - c(\mathbf{0}, w) + c(\mathbf{0}, \mathbf{0}),$$
$$\bar{u}^{\bar{c}}(w) := u^c(w) - u(\mathbf{0}) + c(\mathbf{0}, w) - c(\mathbf{0}, \mathbf{0}).$$

Then $\bar{u}$ is a $\bar{c}$-convex function, $\bar{u}^{\bar{c}}$ is its $\bar{c}$-conjugate, $T_{\bar{u}} = T_u$, and $T_{\bar{u}^{\bar{c}}} = T_{u^c}$, so in particular $(T_{\bar{u}})_\sharp f = g$ and $(T_{\bar{u}^{\bar{c}}})_\sharp g = f$. In addition, because by assumption $\mathbf{0} \in X'$, $\bar{u}$ is twice differentiable at $\mathbf{0}$ and $\bar{u}^{\bar{c}}$ is twice differentiable at $\mathbf{0} = T_{\bar{u}}(\mathbf{0})$. Let us define $P := D^2 \bar{u}(\mathbf{0})$, and $M := D_{xy} \bar{c}(\mathbf{0}, \mathbf{0})$. Then, since $\bar{c}(\cdot, \mathbf{0}) = \bar{c}(\mathbf{0}, \cdot) \equiv 0$ and $\bar{c} \in C^2$, a Taylor expansion gives

$$\bar{u}(z) = \frac{1}{2} P z \cdot z + o(|z|^2), \qquad \bar{c}(z, w) = M z \cdot w + o(|z|^2 + |w|^2).$$

Let us observe that, since by assumption $f$ and $g$ are bounded away from zero and infinity, by (C3) and (1.38) applied to $\bar{u}$ and $\bar{c}$ we get that $\det(P), \det(M) \neq 0$. In addition (1.37) implies that $P$ is a positive definite symmetric matrix. Hence, we can perform a second change of coordinates: $z \mapsto \tilde{z} := P^{1/2}z,\ w \mapsto \tilde{w} := -P^{-1/2}M^*w$ ($M^*$ being the transpose of $M$), so that, in the new variables,

$$\tilde{u}(\tilde{z}) := \bar{u}(z) = \frac{1}{2}|\tilde{z}|^2 + o(|\tilde{z}|^2),$$

$$\tilde{c}(\tilde{z}, \tilde{w}) := \bar{c}(z, w) = -\tilde{z} \cdot \tilde{w} + o(|\tilde{z}|^2 + |\tilde{w}|^2). \tag{6.4}$$

By an easy computation it follows that $(T_{\tilde{u}})_\sharp \tilde{f} = \tilde{g}$, where[2]

$$\tilde{f}(\tilde{z}) := \det(P^{-1/2})\, f(P^{-1/2}\tilde{z}),$$

$$\tilde{g}(\tilde{w}) := \det\big((M^*)^{-1} P^{1/2}\big)\, g((M^*)^{-1} P^{1/2}\tilde{w}). \tag{6.5}$$

Notice that

$$D_{\tilde{z}\tilde{z}}\tilde{c}(0, 0) = D_{\tilde{w}\tilde{w}}\tilde{c}(0, 0) = \mathbf{0}_{n \times n},$$

$$-D_{\tilde{z}\tilde{w}}\tilde{c}(0, 0) = \mathrm{Id}, \tag{6.6}$$

$$D^2\tilde{u}(0) = \mathrm{Id},$$

so, using (1.38), we deduce that

$$\frac{\tilde{f}(0)}{\tilde{g}(0)} = \frac{\det\big(D^2\tilde{u}(0) + D_{\tilde{z}\tilde{z}}\tilde{c}(0, 0)\big)}{\big|\det\big(D_{\tilde{z}\tilde{w}}\tilde{c}(0, 0)\big)\big|} = 1. \tag{6.7}$$

To ensure that we can apply Theorems 6.5 and 6.11, we now perform the following dilation: for $\rho > 0$ we define

$$u_\rho(\tilde{z}) := \frac{1}{\rho^2}\tilde{u}(\rho\tilde{z}), \qquad c_\rho(\tilde{z}, \tilde{w}) := \frac{1}{\rho^2}\tilde{c}(\rho\tilde{z}, \rho\tilde{w}).$$

We claim that, provided $\rho$ is sufficiently small, $u_\rho$ and $c_\rho$ satisfy the assumptions of Theorems 6.5 and 6.11.

Indeed, it is immediate to check that $u_\rho$ is a $c_\rho$-convex function. Also, by the same argument as above, from the relation $(T_{\tilde{u}})_\sharp \tilde{f} = \tilde{g}$ we deduce that $T_{u_\rho}$ sends $\tilde{f}(\rho\tilde{z})$ onto $\tilde{g}(\rho\tilde{w})$. In addition, since we can freely multiply both densities by a same constant, it actually follows from (6.7) that $(T_{u_\rho})_\sharp f_\rho = g_\rho$, where

$$f_\rho(\tilde{z}) := \frac{\tilde{f}(\rho\tilde{z})}{\tilde{f}(0)}, \qquad g_\rho(\tilde{w}) := \frac{\tilde{g}(\rho\tilde{w})}{\tilde{g}(0)}.$$

---

[2] An easy way to check this is to observe that the measures $\mu := f(x)dx$ and $\nu := g(y)dy$ are independent of the choice of coordinates, hence (6.5) follows from the identities

$$f(x)dx = \tilde{f}(\tilde{x})d\tilde{x}, \qquad g(y)dy = \tilde{g}(\tilde{y})d\tilde{y}.$$

In particular, since $f$ and $g$ are continuous, we get

$$|f_\rho - 1| + |g_\rho - 1| \to 0 \qquad \text{inside } B_3 \qquad (6.8)$$

as $\rho \to 0$. Also, by (6.4) we get that, for any $\tilde{z}, \tilde{w} \in B_3$,

$$u_\rho(\tilde{z}) = \frac{1}{2}|\tilde{z}|^2 + o(1), \qquad c_\rho(\tilde{z}, \tilde{w}) = -\tilde{z} \cdot \tilde{w} + o(1), \qquad (6.9)$$

where $o(1) \to 0$ as $\rho \to 0$. In particular, (6.20) and (6.21) hold with any positive constants $\delta_0$, $\eta_0$ provided $\rho$ is small enough.

Furthermore, by the second order differentiability of $\tilde{u}$ at $\mathbf{0}$ it follows that the multivalued map $\tilde{z} \mapsto \partial^- \bar{u}(\tilde{z})$ is differentiable at $\mathbf{0}$ (see Theorem A.5) with gradient equal to the identity matrix (see (6.4)), hence

$$\partial^- u_\rho(\tilde{z}) \subset B_{\gamma_\rho}(\tilde{z}) \qquad \forall \tilde{z} \in B_2,$$

where $\gamma_\rho \to 0$ as $\rho \to 0$. Since $\partial_{c_\rho} u_\rho \subset c_\rho\text{-exp}\left(\partial^- u_\rho\right)$ (by (1.33)) and $\| c_\rho\text{-exp} - \text{Id} \|_\infty = o(1)$ (by (6.9)), we get

$$ilde\partial_{c_\rho} u_\rho(\tilde{z}) \subset B_{\delta_\rho}(\tilde{z}) \qquad \forall \tilde{z} \in B_3, \qquad (6.10)$$

with $\delta_\rho = o(1)$ as $\rho \to 0$. Moreover, the $c_\rho$-conjugate of $u_\rho$ is easily seen to be

$$u_\rho^{c_\rho}(\tilde{w}) = \bar{u}^{\bar{c}}\left(\rho(M^*)^{-1} P^{1/2}\tilde{w}\right).$$

Since $u^c$ is twice differentiable at $\mathbf{0}$, so is $u_\rho^{c_\rho}$. In addition, an easy computation [3] shows that $D^2 u_\rho^{c_\rho}(\mathbf{0}) = \text{Id}$. Hence, arguing as above we obtain that

$$\partial_{c_\rho^*} u_\rho^{c_\rho}(\tilde{w}) \subset B_{\delta_\rho'}(\tilde{w}) \qquad \forall \tilde{w} \in B_3, \qquad (6.11)$$

with $\delta_\rho' = o(1)$ as $\rho \to 0$.

We now define [4]

$$\mathcal{C}_1 := \overline{B}_1, \qquad \mathcal{C}_2 := \partial_{c_\rho} u_\rho(\mathcal{C}_1).$$

---

[3] For instance, this follows by differentiating both relations

$$D_{\tilde{z}} c_\rho\left(\tilde{z}, T_{u_\rho}(\tilde{z})\right) = -\nabla u_\rho(\tilde{z}) \quad \text{and} \quad D_{\tilde{w}} c_\rho\left(T_{u_\rho^{c_\rho}}(\tilde{w}), \tilde{w}\right) = -\nabla u_\rho^{c_\rho}(\tilde{w})$$

at $\mathbf{0}$, and using then (6.6) and the fact that $\nabla T_{u_\rho^{c_\rho}}(\mathbf{0}) = [\nabla T_{u_\rho}(\mathbf{0})]^{-1}$ and $D^2 u_\rho(\mathbf{0}) = \text{Id}$.

[4] We will use the following notation (see Subsection 1.2.2): if $E \subset X$ then

$$\partial_c u(E) := \bigcup_{x \in E} \partial_c u(x), \qquad \partial^- u(E) := \bigcup_{x \in E} \partial^- u(x).$$

Observe that both $\mathcal{C}_1$ and $\mathcal{C}_2$ are closed (since the $c$-subdifferential of a compact set is closed). Also, thanks to (6.10), by choosing $\rho$ small enough we can ensure that $B_{1/3} \subset \mathcal{C}_2 \subset B_3$. Finally, it follows from (1.33) that

$$(T_{u_\rho})^{-1}(\mathcal{C}_2) \setminus \mathcal{C}_1 \subset (T_{u_\rho})^{-1}(\{\text{points of non-differentiability of } u_\rho^{c_\rho}\}),$$

and since this latter set has measure zero, a simple computation shows that

$$\left(T_{u_\rho}\right)_{\sharp}(f_\rho \mathbf{1}_{\mathcal{C}_1}) = g_\rho \mathbf{1}_{\mathcal{C}_2}.$$

Thus, thanks to (6.19), we get that for any $\beta < 1$ the assumptions of Theorem 6.5 are satisfied, provided we choose $\rho$ sufficiently small. Moreover, if in addition $c \in C_{\text{loc}}^{k+2,\alpha}(X \times Y)$, $f \in C_{\text{loc}}^{k,\alpha}(X)$, and $g \in C_{\text{loc}}^{k,\alpha}(Y)$, then also the assumptions of Theorem 6.11 are satisfied.

Hence, by applying Theorem 6.5 (resp. Theorem 6.11) we deduce that $u_\rho \in C^{1,\beta}(B_{1/7})$ (resp. $u_\rho \in C^{k+2,\alpha}(B_{1/9})$), so going back to the original variables we get the existence of a neighborhood $\mathcal{U}_{\bar{x}}$ of $\bar{x}$ such that $u \in C^{1,\beta}(\mathcal{U}_{\bar{x}})$ (resp. $u \in C^{k+2,\alpha}(\mathcal{U}_{\bar{x}})$). This implies in particular that $T_u \in C^{0,\beta}(\mathcal{U}_{\bar{x}})$ (resp. $T_u \in C^{k+1,\alpha}(\mathcal{U}_{\bar{x}})$). Moreover, it follows by Corollary 6.8 that $T_u(\mathcal{U}_{\bar{x}})$ contains a neighborhood of $\bar{y}$.

We now observe that, by symmetry, we can also apply Theorem 6.5 (resp. Theorem 6.11) to $u_\rho^{c_\rho}$. Hence, there exists a neighborhood $\mathcal{V}_{\bar{y}}$ of $\bar{y}$ such that $T_{u^c} \in C^{0,\beta}(\mathcal{V}_{\bar{y}})$. Since $T_u$ and $T_{u^c}$ are inverse to each other (see (6.3)) we deduce that, possibly reducing the size of $\mathcal{U}_{\bar{x}}$, $T_u$ is a homeomorphism (resp. diffeomorphism) between $\mathcal{U}_{\bar{x}}$ and $T_u(\mathcal{U}_{\bar{x}})$. Let us consider the open sets

$$X'' := \bigcup_{\bar{x} \in X'} \mathcal{U}_{\bar{x}}, \qquad Y'' := \bigcup_{\bar{x} \in X'} T_u(\mathcal{U}_{\bar{x}}),$$

and define the (relatively) closed $\Sigma_X := X \setminus X''$, $\Sigma_Y := Y \setminus Y''$. Since $X'' \supset X'$, $X''$ is a set of full measure, so $|\Sigma_X| = 0$. In addition, since $\Sigma_Y = Y \setminus Y'' \subset Y \setminus T_u(X')$ and $T_u(X')$ has full measure in $Y$, we also get that $|\Sigma_Y| = 0$.

Finally, since $T_u : X \setminus \Sigma_X \to Y \setminus \Sigma_Y$ is a local homeomorphism (resp. diffeomorphism), by (6.3) it follows that $T_u : X \setminus \Sigma_X \to Y \setminus \Sigma_Y$ is a global homeomorphism (resp. diffeomorphism), which concludes the proof. $\qquad\square$

*Proof of Theorem* 6.2. The only difference with respect to the situation in Theorem 6.1 is that now the cost function $c = d^2/2$ is not smooth on the whole $M \times M$. However, even if $d^2/2$ is not everywhere smooth and $M$ is not necessarily compact, it is still true that the $c$-convex function $u$

provided by Theorem 1.29 is locally semiconvex (*i.e.*, it is locally semi-convex when seen in any chart) [48, 54]. In addition, as shown in [33, Proposition 4.1] (see also [50, Section 3]), if $u$ is twice differentiable at $x$, then the point $T_u(x)$ is not in the cut-locus of $x$. Since the cut-locus is closed and $d^2/2$ is smooth outside the cut-locus, we deduce the existence of a set $X$ of full measure such that, if $x_0 \in X$, then: (1) $u$ is twice differentiable at $x_0$; (2) there exists a neighborhood $\mathcal{U}_{x_0} \times \mathcal{V}_{T_u(x_0)} \subset M \times M$ of $(x_0, T_u(x_0))$ such that $c \in C^\infty(\mathcal{U}_{x_0} \times \mathcal{V}_{T_u(x_0)})$. Hence, by taking a local chart around $(x_0, T_u(x_0))$, the same proof as the one of Theorem 6.1 shows that $T_u$ is a local homeomorphism (resp. diffeomorphism) around almost every point. Using as before that $T_u : M \to M$ is invertible a.e., it follows that $T_u$ is a global homeomorphism (resp. diffeomorphism) out-side a closed singular set of measure zero. $\qquad\square$

## 6.2. $C^{1,\beta}$ regularity and strict $c$-convexity

In this and the next section we prove that, if in some open set a $c$-convex function $u$ is sufficiently close to a parabola and the cost function is close to the linear one, then $u$ is smooth in some smaller set.

The idea of the proof (which is reminiscent of the argument introduced by Caffarelli in [19] to show $W^{2,p}$ and $C^{2,\alpha}$ estimates for the classical Monge-Ampère equation, though several additional complications arise in our case) is the following: since the cost function is close to the linear one and both densities are almost constant, $u$ is close to a convex func-tion $v$ solving an optimal transport problem with linear cost and constant densities (Lemma 6.3). In addition, since $u$ is close to a parabola, so is $v$. Hence, by [55] and Caffarelli's regularity theory, $v$ is smooth, and we can use this information to deduce that $u$ is even closer to a second parabola (given by the second order Taylor expansion of $v$ at the origin) inside a small neighborhood around of origin. By rescaling back this neighbor-hood at scale 1 and iterating this construction, we obtain that $u$ is $C^{1,\beta}$ at the origin for some $\beta \in (0, 1)$. Since this argument can be applied at every point in a neighborhood of the origin, we deduce that $u$ is $C^{1,\beta}$ there, see Theorem 6.5. (A similar strategy has also been used in [26] to show regularity optimal transport maps for the cost $|x - y|^p$, either when $p$ is close to 2 or when $X$ and $Y$ are sufficiently far from each other.)

Once this result is proved, we know that $\partial^- u$ is a singleton at every point, so it follows from (1.33) that

$$\partial_c u(x) = \text{c-exp}_x(\partial^- u(x)),$$

see Remark 6.6 below. (The above identity is exactly what in general may fail for general $c$-convex functions, unless the MTW condition holds

[78].) Thanks to this fact, we obtain that $u$ enjoys a comparison principle (Proposition 6.10), and this allows us to use a second approximation argument with solutions of the classical Monge-Ampère equation (in the spirit of [19, 69]) to conclude that $u$ is $C^{2,\sigma'}$ in a smaller neighborhood, for some $\sigma' > 0$. Then, higher regularity follows from standard elliptic estimates, see Theorem 6.11.

**Lemma 6.3.** *Let $C_1$ and $C_2$ be two closed sets such that*

$$B_{1/K} \subset C_1, C_2 \subset B_K \tag{6.12}$$

*for some $K \geq 1$, $f$ and $g$ two densities supported respectively in $C_1$ and $C_2$, and $u : C_1 \to \mathbb{R}$ a c-convex function such that $\partial_c u(C_1) \subset B_K$ and $(T_u)_\sharp f = g$. Let $\rho > 0$ be such that $|C_1| = |\rho C_2|$ (where $\rho C_2$ denotes the dilation of $C_2$ with respect to the origin), and let $v$ be a convex function such that $\nabla v_\sharp \mathbf{1}_{C_1} = \mathbf{1}_{\rho C_2}$ and $v(\mathbf{0}) = u(\mathbf{0})$. Then there exists an increasing function $\omega : \mathbb{R}^+ \to \mathbb{R}^+$, depending only $K$, and satisfying $\omega(\delta) \geq \delta$ and $\omega(0^+) = 0$, such that, if*

$$\|f - \mathbf{1}_{C_1}\|_\infty + \|g - \mathbf{1}_{C_2}\|_\infty \leq \delta \tag{6.13}$$

*and*

$$\|c(x, y) + x \cdot y\|_{C^2(B_K \times B_K)} \leq \delta, \tag{6.14}$$

*then*

$$\|u - v\|_{C^0(B_{1/K})} \leq \omega(\delta).$$

*Proof.* Assume the lemma is false. Then there exist $\varepsilon_0 > 0$, a sequence of closed sets $C_1^h, C_2^h$ satisfying (6.12), functions $f_h, g_h$ satisfying (6.13), and costs $c_h$ converging in $C^2$ to $-x \cdot y$, such that

$$u_h(\mathbf{0}) = v_h(\mathbf{0}) = 0 \quad \text{and} \quad \|u_h - v_h\|_{C^0(B_{1/K})} \geq \varepsilon_0,$$

where $u_h$ and $v_h$ are as in the statement. First, we extend $u_h$ an $v_h$ to $B_K$ as

$$u_h(x) := \sup_{z \in C_1^h, y \in \partial_{c_h} u_h(z)} \{u_h(z) - c_h(x, y) + c_h(z, y)\},$$

$$v_h(x) := \sup_{z \in C_1^h, p \in \partial^- v_h(z)} \{u_h(z) + p \cdot (x - z)\}.$$

Notice that, since by assumption $\partial_{c_h} u_h(C_1^h) \subset B_K$, we have $\partial_{c_h} u_h(B_K) \subset B_K$. Also, $(T_{u_h})_\sharp f_h = g_h$ gives that $\int f_h = \int g_h$, so it follows from (6.13) that

$$\rho_h = \left(|C_1^h|/|C_2^h|\right)^{1/n} \to 1 \quad \text{as } h \to \infty,$$

which implies that $\partial^- v_h(B_K) \subset B_{\rho_h K} \subset B_{2K}$ for $h$ large. Thus, since the $C^1$-norm of $c_h$ is uniformly bounded, we deduce that both $u_h$ and $v_h$ are uniformly Lipschitz. Recalling that $u_h(0) = v_h(0) = 0$, we get that, up to a subsequence, $u_h$ and $v_h$ uniformly converge inside $B_K$ to $u_\infty$ and $v_\infty$ respectively, where

$$u_\infty(0) = v_\infty(0) = 0 \quad \text{and} \quad \|u_\infty - v_\infty\|_{C^0(B_{1/K})} \geq \varepsilon_0. \qquad (6.15)$$

In addition $f_h$ (resp. $g_h$) weak-$*$ converge in $L^\infty$ to some density $f_\infty$ (resp. $g_\infty$) supported in $\overline{B}_K$. Also, since $\rho_h \to 1$, using (6.13) we get that $\mathbf{1}_{C_1^h}$ (resp. $\mathbf{1}_{\rho_h C_2^h}$) weak-$*$ converges in $L^\infty$ to $f_\infty$ (resp. $g_\infty$). Finally we remark that, because of (6.13) and the fact that $C_1^h \supset B_{1/K}$, we also have

$$f_\infty \geq \mathbf{1}_{B_{1/K}}.$$

In order to get a contradiction we have to show that $u_\infty = v_\infty$ in $B_{1/K}$. To see this, we apply [95, Theorem 5.20] (see also the proof of Theorem 1.14) to deduce that both $\nabla u_\infty$ and $\nabla v_\infty$ are optimal transport maps for the linear cost $-x \cdot y$ sending $f_\infty$ onto $g_\infty$. By uniqueness of the optimal map we deduce that $\nabla v_\infty = \nabla u_\infty$ almost everywhere inside $B_{1/K} \subset \operatorname{spt} f_\infty$, hence $u_\infty = v_\infty$ in $B_{1/K}$ (since $u_\infty(0) = v_\infty(0) = 0$), contradicting (6.15). $\qquad \square$

Here and in the sequel, we use $\mathcal{N}_r(E)$ to denote the $r$-neighborhood of a set $E$.

**Lemma 6.4.** *Let $u$ and $v$ be, respectively, c-convex and convex, let $D \in \mathbb{R}^{n \times n}$ be a symmetric matrix satisfying*

$$\mathrm{Id}/K \leq D \leq K\,\mathrm{Id} \qquad (6.16)$$

*for some $K \geq 1$, and define the ellipsoid*

$$E(x_0, h) := \left\{ x : D(x - x_0) \cdot (x - x_0) \leq h \right\}, \qquad h > 0.$$

*Assume that there exist small positive constants $\varepsilon, \delta$ such that*

$$\|v - u\|_{C^0(E(x_0,h))} \leq \varepsilon, \qquad \|c + x \cdot y\|_{C^2(E(x_0,h) \times \partial_c u(E(x_0,h)))} \leq \delta. \quad (6.17)$$

*Then*

$$\partial_c u\left(E(x_0, h - \sqrt{\varepsilon})\right) \subset \mathcal{N}_{K'(\delta + \sqrt{h\varepsilon})}\left(\partial v(E(x_0, h))\right) \qquad \forall\, 0 < \varepsilon < h^2 \leq 1, \tag{6.18}$$

*where $K'$ depends only on $K$.*

*Proof.* Up to a change of coordinates we can assume that $x_0 = \mathbf{0}$, and to simplify notation we set $E_h := E(x_0, h)$. Let us define

$$\bar{v}(x) := v(x) + \varepsilon + 2\sqrt{\varepsilon}(Dx \cdot x - h),$$

so that $\bar{v} \geq u$ outside $E_h$, and $\bar{v} \leq u$ inside $E_{h-\sqrt{\varepsilon}}$. Then, taking a $c$-support to $u$ in $E_{h-\sqrt{\varepsilon}}$ (*i.e.*, a function $C_{x,y}$ as in (1.29), with $x \in E_{h-\sqrt{\varepsilon}}$ and $y \in \partial_c u(x)$), moving it down and then lifting it up until it touches $\bar{v}$ from below, we see that it has to touch the graph of $\bar{v}$ at some point $\bar{x} \in E_h$: in other words [5]

$$\partial_c u(E_{h-\sqrt{\varepsilon}}) \subset \partial_c \bar{v}(E_h).$$

By (6.16) we see that diam $E_h \leq \sqrt{Kh}$, so by a simple computation (using again (6.16)) we get

$$\partial^-\bar{v}(E_h) \subset \mathcal{N}_{4K\sqrt{Kh\varepsilon}}\big(\partial^-v(E_h)\big).$$

Thus, since $\partial_c\bar{v}(E_h) \subset \text{c-exp}\big(\partial^-\bar{v}(E_h)\big)$ (by (1.33)) and $\|\text{c-exp} - \text{Id}\|_{C^0} \leq \delta$ (by (6.17)), we easily deduce that

$$\partial_c u(E_{h-\sqrt{\varepsilon}}) \subset \mathcal{N}_{K'(\delta+\sqrt{h\varepsilon})}\big(\partial^-v(E_h)\big),$$

proving (6.18). $\qquad\square$

**Theorem 6.5.** *Let $\mathcal{C}_1$ and $\mathcal{C}_2$ be two closed sets satisfying*

$$B_{1/3} \subset \mathcal{C}_1, \mathcal{C}_2 \subset B_3,$$

*let $f, g$ be two densities supported in $\mathcal{C}_1$ and $\mathcal{C}_2$ respectively, and let $u : \mathcal{C}_1 \to \mathbb{R}$ be a c-convex function such that $\partial_c u(\mathcal{C}_1) \subset B_3$ and $(T_u)_\sharp f = g$. Then, for every $\beta \in (0, 1)$ there exist constants $\delta_0, \eta_0 > 0$ such that the following holds: if*

$$\|f - \mathbf{1}_{\mathcal{C}_1}\|_\infty + \|g - \mathbf{1}_{\mathcal{C}_2}\|_\infty \leq \delta_0, \tag{6.19}$$

$$\|c(x, y) + x \cdot y\|_{C^2(B_3 \times B_3)} \leq \delta_0, \tag{6.20}$$

*and*

$$\left\|u - \frac{1}{2}|x|^2\right\|_{C^0(B_3)} \leq \eta_0, \tag{6.21}$$

*then $u \in C^{1,\beta}(B_{1/7})$.*

---

[5] Even if $\bar{v}$ is not $c$-convex, it still makes sense to consider its $c$-subdifferential (notice that the $c$-subdifferential of $\bar{v}$ may be empty at some points). In particular, the inclusion $\partial_c\bar{v}(x) \subset \text{c-exp}_x(\partial^-\bar{v}(x))$ still holds.

*Proof.* We divide the proof into several steps.

• *Step* 1 : *u is close to a strictly convex solution of the Monge Ampère equation.* Let $v : \mathbb{R}^n \to \mathbb{R}$ be a convex function such that $\nabla v_{\sharp} \mathbf{1}_{\mathcal{C}_1} = \mathbf{1}_{\rho \mathcal{C}_2}$ with $\rho = (|\mathcal{C}_1|/|\mathcal{C}_2|)^{1/n}$. Up to a adding a constant to $v$, without loss of generality we can assume that $v(0) = u(0)$. Hence, we can apply Lemma 6.3 to obtain

$$\|v - u\|_{C^0(B_{1/3})} \leq \omega(\delta_0) \tag{6.22}$$

for some (universal) modulus of continuity $\omega : \mathbb{R}^+ \to \mathbb{R}^+$, which combined with (6.21) gives

$$\left\| v - \frac{1}{2}|x|^2 \right\|_{C^0(B_{1/3})} \leq \eta_0 + \omega(\delta_0). \tag{6.23}$$

Also, since $\int_{\mathcal{C}_1} f = \int_{\mathcal{C}_2} g$, it follows easily from (6.19) that $|\rho - 1| \leq 3\delta_0$. By these two facts we get that $\partial^- v(B_{7/24}) \subset B_{15/48} \subset \rho \mathcal{C}_2$ provided $\delta_0$ and $\eta_0$ are small enough (recall that $v$ is convex and that $B_{1/3} \subset \mathcal{C}_2$), so, thanks to Remark 1.24, $v$ is a convex Aleksandrov solution to the Monge-Ampère equation

$$\det D^2 v = 1 \quad \text{in } B_{7/24}. \tag{6.24}$$

In addition, by (6.23) and Theorem 2.9 we see that, for $\delta_0$, $\eta_0$ small enough, $v$ is strictly convex in $B_{1/4}$. A simple compactness argument shows that we the modulus of strict convexity of $v$ inside $B_{1/4}$ is universal [6]. So, by classical Pogorelov and Schauder estimates (see Section 2.3), we obtain the existence of a universal constant $K_0 \geq 1$ such that

$$\|v\|_{C^3(B_{1/5})} \leq K_0, \qquad \mathrm{Id}/K_0 \leq D^2 v \leq K_0 \, \mathrm{Id} \quad \text{in } B_{1/5}. \tag{6.25}$$

In particular, there exists a universal value $\bar{h} > 0$ such that, for all $x \in B_{1/7}$,

$$Q(x, v, h) := \left\{ z : v(z) \leq v(x) + \nabla v(x) \cdot (z - x) + h \right\} \subset\subset B_{1/6} \quad \forall h \leq \bar{h}.$$

---

[6] To see this, suppose there exists a sequence of strictly convex functions $v_k$ satisfying (6.24) and (6.23), but whose modulus of strict convexity inside $B_{1/4}$ is going to 0. By compactness, we can find a limiting function $v_\infty$ satisfying both (6.24) and (6.23) which is not strictly convex in $B_{1/4}$ (see Lemma 2.10).Hence, there exists a supporting hyperplane $\ell_\infty$ to $v_\infty$ such that the set $W := \{v_\infty = \ell_\infty\}$ intersects $B_{1/4}$ and it is not reduced to a point. Then, by Theorem 2.9 we know that $W$ has to cross $\partial B_{7/24}$, but this is impossible if $\eta_0$ and $\delta_0$ are sufficiently small.

• *Step 2* : *Sections of u are close to sections of v.* Given $x \in B_{1/7}$ and $y \in \partial_c u(x)$, we define

$$S(x, y, u, h) := \big\{ z : \ u(z) \leq u(x) - c(z, y) + c(x, y) + h \big\}.$$

We claim that, if $\delta_0$ is small enough, then for all $x \in B_{1/7}$, $y \in \partial_c u(x)$, and $h \leq \bar{h}/2$, it holds

$$Q(x, v, h - K_1 \sqrt{\omega(\delta_0)}) \subset S(x, y, u, h) \subset Q(x, v, h + K_1 \sqrt{\omega(\delta_0)}),$$
(6.26)

where $K_1 > 0$ is a universal constant.

Let us show the first inclusion. For this, take $x \in B_{1/7}$, $y \in \partial_c u(x)$, and define

$$p_x := -D_x c(x, y) \in \partial^- u(x).$$

Since $v$ has universal $C^2$ bounds (see (6.25)) and $u$ is semi-convex (with a universal bound), a simple interpolation argument gives

$$|p_x - \nabla v(x)| \leq K' \sqrt{\|u - v\|_{C^0(B_{1/5})}} \leq K' \sqrt{\omega(\delta_0)} \quad \forall x \in B_{1/7}. \quad (6.27)$$

In addition, by (6.20),

$$|y - p_x| \leq \|D_x c + \mathrm{Id}\|_{C^0(B_3 \times B_3)} \leq \delta_0, \quad (6.28)$$

hence

$$|z \cdot p_x + c(z, y)| \leq |z \cdot p_x - z \cdot y| + |z \cdot y + c(z, y)|$$
$$\leq 2\delta_0 \quad \forall x, z \in B_{1/7}. \quad (6.29)$$

Thus, if $z \in Q(x, v, h - K_1 \sqrt{\omega(\delta_0)})$, by (6.22), (6.27), and (6.29) we get

$$u(z) \leq v(z) + \omega(\delta_0)$$
$$\leq v(x) + \nabla v(x) \cdot (z - x) + h - K_1 \sqrt{\omega(\delta_0)} + \omega(\delta_0)$$
$$\leq u(x) + p_x \cdot z - p_x \cdot x + h - K_1 \sqrt{\omega(\delta_0)}$$
$$\quad + 2\omega(\delta_0) + 2K' \sqrt{\omega(\delta_0)}$$
$$\leq u(x) - c(z, y) + c(x, y) + h - K_1 \sqrt{\omega(\delta_0)}$$
$$\quad + 2\omega(\delta_0) + 2K' \sqrt{\omega(\delta_0)} + 4\delta_0$$
$$\leq u(x) - c(z, y) + c(x, y) + h,$$

provided $K_1 > 0$ is sufficiently large. This proves the first inclusion, and the second one is analogous.

• *Step* 3 : *Both the sections of u and their images are close to ellipsoids with controlled eccentricity, and u is close to a smooth function near* $x_0$. We claim that there exists a universal constant $K_2 \geq 1$ such that the following holds: For every $\eta_0 > 0$ small, there exist small positive constants $h_0 = h_0(\eta_0)$ and $\delta_0 = \delta_0(h_0, \eta_0)$ such that, for all $x_0 \in B_{1/7}$, there is a symmetric matrix $A$ satisfying

$$\mathrm{Id}/K_2 \leq A \leq K_2\,\mathrm{Id}, \qquad \det(A) = 1, \qquad (6.30)$$

and such that, for all $y_0 \in \partial_c u(x_0)$,

$$A\left(B_{\sqrt{h_0/8}}(x_0)\right) \subset S(x_0, y_0, u, h_0) \subset A\left(B_{\sqrt{8h_0}}(x_0)\right),$$
$$A^{-1}\left(B_{\sqrt{h_0/8}}(y_0)\right) \subset \partial_c u(S(x_0, y_0, u, h_0)) \subset A^{-1}\left(B_{\sqrt{8h_0}}(y_0)\right). \qquad (6.31)$$

Moreover

$$\left\| u - C_{x_0, y_0} - \frac{1}{2}\left|A^{-1}(x - x_0)\right|^2 \right\|_{C^0\left(A\left(B_{\sqrt{8h_0}}(x_0)\right)\right)} \leq \eta_0 h_0, \qquad (6.32)$$

where $C_{x_0 y_0}$ is a c-support function for $u$ at $x_0$, see (1.29).

In order to prove the claim, take $h_0 \ll \bar{h}$ small (to be fixed) and $\delta_0 \ll h_0$ such that $K_1\sqrt{\omega(\delta_0)} \leq h_0/2$, where $K_1$ is as in Step 2, so that

$$Q(x_0, v, h_0/2) \subset S(x_0, y_0, u, h_0) \subset Q(x_0, v, 3h_0/2) \subset\subset B_{1/6}. \qquad (6.33)$$

By (6.25) and Taylor formula we get

$$v(x) = v(x_0) + \nabla v(x_0) \cdot (x - x_0)$$
$$+ \frac{1}{2}D^2 v(x_0)(x - x_0) \cdot (x - x_0) + O(|x - x_0|^3), \qquad (6.34)$$

so that defining

$$E(x_0, h_0) := \left\{ x : \frac{1}{2}D^2 v(x_0)(x - x_0) \cdot (x - x_0) \leq h_0 \right\} \qquad (6.35)$$

and using (6.25), we deduce that, for every $h_0$ universally small,

$$E(x_0, h_0/2) \subset Q(x_0, v, h_0) \subset E(x_0, 2h_0). \qquad (6.36)$$

Moreover, always for $h_0$ small, thanks to (6.34) and the uniform convexity of $v$

$$\nabla v\left(E(x_0, h_0)\right) \subset E^*(\nabla v(x_0), 2h_0) \subset \nabla v\left(E(x_0, 3h_0)\right) \qquad (6.37)$$

where we have set

$$E^*(\bar{y}, h_0) := \left\{ y : \frac{1}{2}[D^2 v(\bar{y})]^{-1}(y - \bar{y}) \cdot (y - \bar{y}) \le h_0 \right\}.$$

By Lemma 6.4, (6.36), and (6.37) applied with $3h_0$ in place of $h_0$, we deduce that for $\delta_0 \ll h_0 \ll \bar{h}$

$$\partial_c u\big(S(x_0, y_0, u, h_0)\big) \subset \mathcal{N}_{K'' \sqrt{\omega(\delta_0)}}(\nabla v(E(x_0, 3h_0))) \\ \subset E^*(\nabla v(x_0), 7h_0). \tag{6.38}$$

Moreover, by (6.27), if $y_0 \in \partial_c u(x_0)$ and we set $p_{x_0} := -D_x c(x_0, y_0)$, then

$$|y_0 - \nabla v(x_0)| \le |p_{x_0} - \nabla v(x_0)| + \|D_x c + \mathrm{Id}\|_{C^0(B_3 \times B_3)} \le \sqrt{\omega(\delta_0)} + \delta_0.$$

Thus, choosing $\delta_0$ sufficiently small, it holds

$$E^*(\nabla v(x_0), 7h_0) \subset E^*(y_0, 8h_0) \quad \forall y_0 \in \partial_c u(x_0). \tag{6.39}$$

We now want to show that

$$E^*(y_0, h_0/8) \subset \partial_c u\big(S(x_0, y_0, u, h_0)\big) \quad \forall y_0 \in \partial_c u(x_0).$$

Observe that, arguing as above, we get

$$E^*(y_0, h_0/8) \subset E^*(\nabla v(x_0), h_0/7) \quad \forall y_0 \in \partial_c u(x_0) \tag{6.40}$$

provided $\delta_0$ is small enough, so it is enough to prove that

$$E^*(\nabla v(x_0), h_0/7) \subset \partial_c u\big(S(x_0, y_0, u, h_0)\big).$$

For this, let us define the $c^*$-convex function $u^c : B_3 \to \mathbb{R}$ and the convex function $v^* : B_3 \to \mathbb{R}$ as

$$u^c(y) := \sup_{x \in B_{1/5}} \big\{ -c(x, y) - u(x) \big\}, \qquad v^*(y) := \sup_{x \in B_{1/5}} \big\{ x \cdot y - v(x) \big\}$$

(see (6.1)). Then it is immediate to check that

$$|u^c - v^*| \le \omega(\delta_0) + \delta_0 \le 2\omega(\delta_0) \qquad \text{on } B_3. \tag{6.41}$$

Also, in view of (6.25), $v^*$ is a uniformly convex function of class $C^3$ on the open set $\nabla v(B_{1/5})$. In addition, since

$$F \subset \partial_c u(\partial_{c^*} u^c(F)) \qquad \text{for any set } F, \tag{6.42}$$

thanks to (6.33) and (6.36) it is enough to show

$$\partial_{c^*} u^c(E^*(\nabla v(x_0), h_0/7)) \subset E(x_0, h_0/4). \qquad (6.43)$$

For this, we apply Lemma 6.4 to $u^c$ and $v^*$ to infer

$$\partial_{c^*} u^c(E^*(\nabla v(x_0), h_0/7)) \subset \mathcal{N}_{K''' \sqrt{\omega(\delta)}}\big(\nabla v^*(E^*(\nabla v(x_0), h_0/7))\big)$$
$$\subset E(x_0, h_0/4),$$

where we used that

$$\nabla v^* = [\nabla v]^{-1} \quad \text{and} \quad D^2 v^*(\nabla v(x_0)) = [D^2 v(x_0)]^{-1}.$$

Thus, recalling (6.38), we have proved that there exist $h_0$ universally small, and $\delta_0$ small depending on $h_0$, such that

$$E^*(\nabla v(x_0), h_0/7) \subset \partial_c u(S(x_0, y_0, u, h_0))$$
$$\subset E^*(\nabla v(x_0), 7h_0) \qquad \forall x_0 \in B_{1/7}. \qquad (6.44)$$

Using (6.33), (6.36), (6.39), and (6.40), this proves (6.31) with $A :=$ $[D^2 v(x_0)]^{-1/2}$. Also, thanks to (6.24) and (6.25), (6.30) holds.

In order to prove the second part of the claim, we exploit (6.22), (6.20), (6.28), (6.27), (6.34), and (6.30) (recall that $C_{x_0, y_0}$ is defined in (1.29) and that $A = [D^2 v(x_0)]^{-1/2}$):

$$\left\| u - C_{x_0, y_0} - \frac{1}{2}|A^{-1}(x - x_0)|^2 \right\|_{C^0(E(x_0, 8h_0))}$$

$$= \left\| u - C_{x_0, y_0} - \frac{1}{2}D^2 v(x_0)(x - x_0) \cdot (x - x_0) \right\|_{C^0(E(x_0, 8h_0))}$$

$$\leq 2\|u - v\|_{C^0(E(x_0, 8h_0))} + \|c(x, y_0) + x \cdot y_0\|_{C^0(E(x_0, 8h_0))}$$
$$+ \|c(x_0, y_0) + x_0 \cdot y_0\|_{C^0(E(x_0, 8h_0))} + \left\| (y_0 - p_{x_0}) \cdot (x - x_0) \right\|_{C^0(E(x_0, 8h_0))}$$
$$+ \left\| (p_{x_0} - \nabla v(x_0)) \cdot (x - x_0) \right\|_{C^0(E(x_0, 8h_0))}$$
$$+ \left\| v - v(x_0) - \nabla v(x_0) \cdot (x - x_0) \right.$$

$$\left. - \frac{1}{2}D^2 v(x_0)(x - x_0) \cdot (x - x_0) \right\|_{C^0(E(x_0, 8h_0))}$$

$$\leq 2\omega(\delta_0) + 3\delta_0 + K' \sqrt{\omega(\delta_0)} + K \left( K_2 \sqrt{8h_0} \right)^3 \leq \eta_0 h_0,$$

where the last inequality follows by choosing first $h_0$ sufficiently small, and then $\delta_0$ much smaller than $h_0$.

• *Step* 4 : *A first change of variables.* Fix $x_0 \in B_{1/7}$, $y_0 \in \partial_c u(x_0)$, define $M := -D_{xy}c(x_0, y_0)$, and consider the change of variables

$$\begin{cases} \bar{x} := x - x_0 \\ \bar{y} := M^{-1}(y - y_0). \end{cases}$$

Notice that, by (6.20), it follows that

$$|M - \mathrm{Id}| + |M^{-1} - \mathrm{Id}| \le 3\delta_0 \tag{6.45}$$

for $\delta_0$ sufficiently small. We also define

$$\bar{c}(\bar{x}, \bar{y}) := c(x, y) - c(x, y_0) - c(x_0, y) + c(x_0, y_0),$$
$$\bar{u}(\bar{x}) := u(x) - u(x_0) + c(x, y_0) - c(x_0, y_0),$$
$$\bar{u}^{\bar{c}}(\bar{y}) := u^c(y) - u^c(y_0) + c(x_0, y) - c(x_0, y_0).$$

Then $\bar{u}$ is $\bar{c}$-convex, $\bar{u}^{\bar{c}}$ is $\bar{c}^*$-convex (where $\bar{c}^*(\bar{y}, \bar{x}) = \bar{c}(\bar{x}, \bar{y})$), and

$$\bar{c}(\cdot, \mathbf{0}) = \bar{c}(\mathbf{0}, \cdot) \equiv 0, \qquad D_{\bar{x}\bar{y}}\bar{c}(\mathbf{0}, \mathbf{0}) = -\mathrm{Id}. \tag{6.46}$$

We also notice that

$$\partial_{\bar{c}}\bar{u}(\bar{x}) = M^{-1}\big(\partial_c u(\bar{x} + x_0) - y_0\big). \tag{6.47}$$

Thus, recalling (6.31), and using (6.45) and (6.47), for $\delta_0$ sufficiently small we obtain

$$A\left(B_{\sqrt{h_0/9}}\right) \quad \subset S(\mathbf{0}, \mathbf{0}, \bar{u}, h_0) \tag{6.48}$$
$$\subset \partial_{\bar{c}^*}\bar{u}^{\bar{c}}\big(\mathrm{co}\left[\partial_{\bar{c}}\bar{u}(S(\mathbf{0}, \mathbf{0}, \bar{u}, h_0))\right]\big) \subset A\left(B_{\sqrt{9h_0}}\right),$$
$$A^{-1}\left(B_{\sqrt{h_0/9}}\right) \quad \subset M^{-1}A^{-1}\left(B_{\sqrt{h_0/8}}\right)$$
$$\subset \mathrm{co}\left[\partial_{\bar{c}}\bar{u}(S(\mathbf{0}, \mathbf{0}, \bar{u}, h_0))\right]$$
$$\subset M^{-1}A^{-1}\left(B_{\sqrt{8h_0}}\right) \subset A^{-1}\left(B_{\sqrt{9h_0}}\right).$$

Since $(T_u)_\sharp f = g$, it follows that $T_{\bar{u}} = \bar{c}\text{-}\exp(\nabla\bar{u})$ satisfies

$$(T_{\bar{u}})_\sharp \bar{f} = \bar{g}, \quad \text{with} \quad \bar{f}(\bar{x}) := f(\bar{x}+x_0), \quad \bar{g}(\bar{y}) := \det(M)\,g(M\bar{y}+y_0)$$

(see for instance the footnote in the proof of Theorem 6.1). Notice that, since $|M - \mathrm{Id}| \le \delta_0$ (by (6.20)), we have $|\det(M) - 1| \le (1 + 2n)\delta_0$ (for $\delta_0$ small), so by (6.19) we get

$$\|\bar{f} - \mathbf{1}_{C_1-x_0}\|_\infty + \|\bar{g} - \mathbf{1}_{M^{-1}(C_2-y_0)}\|_\infty \le 2(1 + n)\delta_0. \tag{6.49}$$

• *Step 5 : A second change of variables and the iteration argument.* We now perform a second change of variable: we set

$$\begin{cases} \tilde{x} := \dfrac{1}{\sqrt{h_0}} A^{-1} \bar{x} \\ \tilde{y} := \dfrac{1}{\sqrt{h_0}} A \bar{y}, \end{cases} \tag{6.50}$$

and define

$$c_1(\tilde{x}, \tilde{y}) := \frac{1}{h_0} \bar{c}\big(\sqrt{h_0} A \tilde{x}, \sqrt{h_0} A^{-1} \tilde{y}\big),$$

$$u_1(\tilde{x}) := \frac{1}{h_0} \bar{u}\big(\sqrt{h_0} A \tilde{x}\big),$$

$$u_1^{c_1}(\tilde{y}) := \frac{1}{h_0} \bar{u}^{\bar{c}}\big(\sqrt{h_0} A^{-1} \tilde{y}\big).$$

We also define

$$f_1(\tilde{x}) := \bar{f}(\sqrt{h_0} A \tilde{x}), \qquad g_1(\tilde{y}) := \bar{g}(\sqrt{h_0} A^{-1} \tilde{y}).$$

Since $\det(A) = 1$ (see (6.30)), it is easy to check that $(T_{u_1})_\sharp f_1 = g_1$ (see the footnote in the proof of Theorem 6.1).
Also, since $\big(\|A\| + \|A^{-1}\|\big) \sqrt{h_0} \ll 1$, it follows from (6.49) that

$$|f_1 - 1| + |g_1 - 1| \le 2(1 + n)\delta_0 \qquad \text{inside } B_3. \tag{6.51}$$

Moreover, defining

$$\mathcal{C}_1^{(1)} := S(\mathbf{0}, \mathbf{0}, u_1, 1), \qquad \mathcal{C}_2^{(1)} := \partial_{c_1} u_1(S(\mathbf{0}, \mathbf{0}, u_1, 1)),$$

both $\mathcal{C}_1^{(1)}$ and $\mathcal{C}_2^{(1)}$ are closed, and thanks to (6.48)

$$B_{1/3} \subset \mathcal{C}_1^{(1)}, \mathcal{C}_2^{(1)} \subset B_3. \tag{6.52}$$

Also, since $(T_{u_1})_\sharp f_1 = g_1$, arguing as in the proof of Theorem 6.1 we get

$$(T_{u_1})_\sharp \big(f_1 \mathbf{1}_{\mathcal{C}_1^{(1)}}\big) = \big(g_1 \mathbf{1}_{\mathcal{C}_2^{(1)}}\big),$$

and by (6.51)

$$\|f_1 \mathbf{1}_{\mathcal{C}_1^{(1)}} - \mathbf{1}_{\mathcal{C}_1^{(1)}}\|_\infty + \|g_1 \mathbf{1}_{\mathcal{C}_2^{(1)}} - \mathbf{1}_{\mathcal{C}_2^{(1)}}\|_\infty \le 2(1 + n)\delta_0.$$

Finally, by (6.46) and (6.32), it is easy to check that

$$\|c_1(\tilde{x}, \tilde{y}) + \tilde{x} \cdot \tilde{y}\|_{C^2(B_3 \times B_3)} \le \delta_0, \qquad \left\| u_1 - \frac{1}{2}|\tilde{x}|^2 \right\|_{C^0(B_3)} \le \eta_0.$$

This shows that $u_1$ satisfies the same assumptions as $u$ with $\delta_0$ replaced by $2(1 + n)\delta_0$. Hence, taking $\delta_0$ slightly smaller, we can apply Step 3 to $u_1$, and to find a symmetric matrix $A_1$ satisfying

$$\text{Id}/K_2 \le A_1 \le K_2 \,\text{Id}, \qquad \det(A_1) = 1,$$

$$A_1\left(B_{\sqrt{h_0/8}}\right) \subset S(\mathbf{0}, \mathbf{0}, u_1, h_0) \subset A_1\left(B_{\sqrt{8h_0}}\right),$$

$$A_1^{-1}\left(B_{\sqrt{h_0/8}}\right) \subset \partial_{c_1}u_1(S(\mathbf{0}, \mathbf{0}, u_1, h_0)) \subset A_1^{-1}\left(B_{\sqrt{8h_0}}\right),$$

$$\left\| u_1 - \frac{1}{2}\left|A_1^{-1}\tilde{x}\right|^2 \right\|_{C^0(A_1(B(0,\sqrt{8h_0})))} \le \eta_0 h_0.$$

(Here $K_2$ and $h_0$ are as in Step 3.)

This allows us to apply to $u_1$ the very same construction as the one used above to define $u_1$ from $\bar{u}$: we set

$$c_2(\tilde{x}, \tilde{y}) := \frac{1}{h_0}c_1\left(\sqrt{h_0}A_1\tilde{x}, \sqrt{h_0}A_1^{-1}\tilde{y}\right), \qquad u_2(\tilde{x}) := \frac{1}{h_0}u_1(\sqrt{h_0}A_1\tilde{x}),$$

so that $(T_{u_2})_\sharp f_2 = g_2$ with

$$f_2(\tilde{x}) := f_1(\sqrt{h_0}A_1\tilde{x}), \qquad g_2(\tilde{y}) := \bar{g}(\sqrt{h_0}A_1^{-1}\tilde{y}).$$

Arguing as before, it is easy to check that $u_2, c_2, f_2, g_2$ satisfy the same assumptions as $u_1, c_1, f_1, g_1$ with exactly the same constants.

So we can keep iterating this construction, defining for any $k \in \mathbb{N}$

$$c_{k+1}(\tilde{x}, \tilde{y}) := \frac{1}{h_0}c_k\left(\sqrt{h_0}A_k\tilde{x}, \sqrt{h_0}A_k^{-1}\tilde{y}\right), \qquad u_{k+1}(\tilde{x}) := \frac{1}{h_0}u_k(\sqrt{h_0}A_k\tilde{x}),$$

where $A_k$ is the matrix constructed in the $k$-th iteration. In this way, if we set

$$M_k := A_k \cdot \ldots \cdot A_1, \qquad \forall k \ge 1,$$

we obtain a sequence of symmetric matrices satisfying

$$\text{Id}/K_2^k \le M_k \le K_2^k \,\text{Id}, \qquad \det(M_k) = 1, \tag{6.53}$$

and such that

$$M_k\left(B_{(h_0/8)^{k/2}}\right) \subset S(\mathbf{0}, \mathbf{0}, u_k, h_0^k) \subset M_k\left(B_{(8h_0)^{k/2}}\right). \tag{6.54}$$

• *Step 6 : $C^{1,\beta}$ regularity.* We now show that, for any $\beta \in (0, 1)$, we can choose $h_0$ and $\delta_0 = \delta_0(h_0)$ small enough so that $u_1$ is $C^{1,\beta}$ at the origin (here $u_1$ is the function constructed in the previous step). This will imply that $u$ is $C^{1,\beta}$ at $x_0$ with universal bounds, which by the arbitrariness of $x_0 \in B_{1/7}$ gives $u \in C^{1,\beta}(B_{1/7})$.

Fix $\beta \in (0, 1)$. Then by (6.53) and (6.54) we get

$$B_{\left(\sqrt{h_0}/(\sqrt{8}K_2)\right)^k} \subset S(0, 0, u_1, h_0^k) \subset B_{\left(K_2\sqrt{8h_0}\right)^k}, \qquad (6.55)$$

so defining $r_0 := \sqrt{h_0}/(\sqrt{8}K_2)$ we obtain

$$\|u_1\|_{C^0(B_{r_0^k})} \le h_0^k = \left(\sqrt{8}K_2 r_0\right)^{2k} \le r_0^{(1+\beta)k},$$

provided $h_0$ (and so $r_0$) is sufficiently small. This implies the $C^{1,\beta}$ regularity of $u_1$ at $\mathbf{0}$, concluding the proof.  □

**Remark 6.6 (Local to global principle).** If $u$ is differentiable at $x$ and $c$ satisfies **(C0)-(C1)**, then every "local $c$-support" at $x$ it is also a "global support" at $x$, that is, $\partial_c u(x) = \text{c-exp}_x(\partial^- u(x))$. To see this, just notice that

$$\emptyset \ne \partial_c u(x) \subset \text{c-exp}_x(\partial^- u(x)) = \{\text{c-exp}_x(\nabla u(x))\}$$

(recall (1.33)), so necessarily the two sets have to coincide.

**Corollary 6.7.** *Let $u$ be as in Theorem* 6.5. *Then $u$ is strictly $c$-convex in $B_{1/7}$. More precisely, for every $\gamma > 2$ there exist $\eta_0, \delta_0 > 0$ depending only on $\gamma$ such that, if the hypothesis of Theorem* 6.5 *are satisfied, then, for all $x_0 \in B_{1/7}$, $y_0 \in \partial_c u(x_0)$, and $C_{x_0,y_0}$ as in* (1.29), *we have*

$$\inf_{\partial B_r(x_0)} \left\{ u - C_{x_0,y_0} \right\} \ge c_0 r^\gamma \qquad \forall\, r \le \text{dist}(x_0, \partial B_{1/7}), \qquad (6.56)$$

*with $c_0 > 0$ universal.*

*Proof.* With the same notation as in the proof of Theorem 6.5, it is enough to show that

$$\inf_{\partial B_r} u_1 \ge r^{1/\beta},$$

where $u_1$ is the function constructed in Step 5 of the proof of Theorem 6.5. Defining $\varrho_0 := K_2\sqrt{8h_0}$, it follows from (6.55) that

$$\inf_{\partial B_{\varrho_0^k}} u_1 \ge h_0^k = \left(\varrho_0/(\sqrt{8}K_2)\right)^{2k} \ge \varrho_0^{\gamma k},$$

provided $h_0$ is small enough.  □

A simple consequence of the above results is the following:

**Corollary 6.8.** *Let $u$ be as in Theorem 6.5, then $T_u(B_{1/7})$ is open.*

*Proof.* Since $u \in C^{1,\beta}(B_{1/7})$ we have that $T_u(B_{1/7}) = \partial_c u(B_{1/7})$ (see Remark 6.6). We claim that it is enough to show that if $y_0 \in \partial_c u(B_{1/7})$, then there exists $\varepsilon = \varepsilon(y_0) > 0$ small such that, for all $|y - y_0| < \varepsilon$, the function $u(\cdot) + c(\cdot, y)$ has a local minimum at some point $\bar{x} \in B_{1/7}$. Indeed, if this is the case, then

$$\nabla u(\bar{x}) = -D_x c(\bar{x}, y),$$

and so $y \in \partial_c u(\bar{x})$ (by Remark 6.6), hence $B_\varepsilon(y_0) \subset T_u(B_{1/7})$.

To prove the above fact, fix $r > 0$ such that $B_r(x_0) \subset B_{1/7}$, and pick $\bar{x}$ a point in $\overline{B}_r(x_0)$ where the function $u(\cdot) + c(\cdot, y)$ attains its minimum, *i.e.*,

$$\bar{x} \in \underset{\overline{B}_r(x_0)}{\operatorname{argmin}} \big\{ u(x) + c(x, y) \big\}.$$

Since, by (6.56),

$$\min_{x \in \partial B_r(x_0)} \big\{ u(x) + c(x, y) \big\} \geq \min_{x \in \partial B_r(x_0)} \big\{ u(x) + c(x, y_0) \big\} - \varepsilon \|c\|_{C^1}$$

$$\geq u(x_0) + c(x_0, y_0) + c_0 r^\gamma - \varepsilon \|c\|_{C^1},$$

while

$$u(x_0) + c(x_0, y) \leq c(x_0, y_0) + u(x_0) + \varepsilon \|c\|_{C^1},$$

choosing $\varepsilon < \frac{c_0}{2\|c\|_{C^1}} r^\gamma$ we obtain that $\bar{x} \in B_r(x_0) \subset B_{1/7}$. This implies that $\bar{x}$ is a local minimum for $u(\cdot) + c(\cdot, y)$, concluding the proof.  $\square$

## 6.3. Comparison principle and $C^{2,\alpha}$ regularity

**Lemma 6.9.** *Let $\Omega$ be an open set, $v \in C^2(\Omega)$, and assume that $\nabla v(\Omega) \subset$ Dom c-exp and that*

$$D^2 v(x) + D_{xx} c\big(x, \text{c-exp}_x(\nabla v(x))\big) \geq 0 \quad \forall x \in \Omega.$$

*Then, for every Borel set $A \subset \Omega$,*

$$|\text{c-exp}(\nabla v(A))| \leq \int_A \frac{\det \big( D^2 v(x) + D_{xx} c\big(x, \text{c-exp}_x(\nabla v(x))\big) \big)}{\big| \det\big( D_{xy} c\big(x, \text{c-exp}_x(\nabla v(x))\big) \big) \big|} \, dx.$$

*In addition, if the map $x \mapsto \text{c-exp}_x(\nabla v(x))$ is injective, then equality holds.*

*Proof.* The proof follows from a direct application of the Area Formula (1.15) once one notices that, differentiating the identity (see (1.32))

$$\nabla v(x) = -D_x c\big(x, \text{c-exp}_x(\nabla v(x))\big),$$

the Jacobian determinant of the $C^1$ map $x \mapsto \text{c-exp}_x(\nabla v(x))$ is given precisely by

$$\frac{\det\big(D^2 v(x) + D_{xx} c\big(x, \text{c-exp}_x(\nabla v(x))\big)\big)}{\big|\det\big(D_{xy} c\big(x, \text{c-exp}_x(\nabla v(x))\big)\big)\big|}. \qquad \square$$

In the next proposition we show a comparison principle between $C^1$ $c$-convex functions and smooth solutions to the Monge-Ampère equation. As already mentioned at the beginning of Section 6.2 (see also Remark 6.6), the $C^1$ regularity of $u$ is crucial to ensure that the $c$-subdifferential coincides with its local counterpart c-exp$(\partial^- u)$.

Given a set $E$, we denote by co$[E]$ its convex hull.

**Proposition 6.10 (Comparison principle).** *Let $u$ be a $c$-convex function of class $C^1$ inside the set $S := \{u < 1\}$, and assume that $u(0) = 0$, $B_{1/K} \subset S \subset B_K$, and that $\nabla u(S) \Subset \text{Dom c-exp}$. Let $f, g$ be two densities such that*

$$\|f/\lambda_1 - 1\|_{C^0(S)} + \|g/\lambda_2 - 1\|_{C^0(T_u(S))} \le \varepsilon \qquad (6.57)$$

*for some constants $\lambda_1, \lambda_2 \in (1/2, 2)$ and $\varepsilon \in (0, 1/4)$, and assume that $(T_u)_\sharp f = g$. Furthermore, suppose that*

$$\|c + x \cdot y\|_{C^2(B_K \times B_K)} \le \delta. \qquad (6.58)$$

*Then there exist a universal constant $\gamma \in (0, 1)$, and $\delta_1 = \delta_1(K) > 0$ small, such that the following holds: Let $v$ be the solution of*

$$\begin{cases} \det(D^2 v) = \lambda_1/\lambda_2 & \text{in } \mathcal{N}_{\delta^\gamma}(\text{co}[S]), \\ v = 1 & \text{on } \partial\big(\mathcal{N}_{\delta^\gamma}(\text{co}[S])\big). \end{cases}$$

*Then*

$$\|u - v\|_{C^0(S)} \le C_K \big(\varepsilon + \delta^{\gamma/n}\big) \qquad \text{provided } \delta \le \delta_1, \qquad (6.59)$$

*where $C_K$ is a constant independent of $\lambda_1, \lambda_2, \varepsilon$, and $\delta$ (but which depends on $K$).*

*Proof.* First of all we observe that, since $u(\mathbf{0}) = 0$, $u = 1$ on $\partial S$, $S \subset B_K$, and $\|c + x \cdot y\|_{C^2(B_K)} \leq \delta \ll 1$, it is easy to check that there exists a universal constant $a_1 > 0$ such that

$$|D_x c(x, y)| \geq a_1 \qquad \forall x \in \partial S, \ y = \text{c-exp}_x(\nabla u(x)). \qquad (6.60)$$

Thanks to (6.60) and (6.58), it follows from the Implicit Function Theorem that, for each $x \in \partial S$, the boundary of the set

$$E_x := \{z \in B_K : c(z, y) - c(x, y) + u(x) \leq 1\}$$

is of class $C^2$ inside $B_K$, and its second fundamental form is bounded by $C_K \delta$, where $C_K > 0$ depends only on $K$. Hence, since $S$ can be written as

$$S := \bigcap_{x \in \partial S} E_x,$$

it follows that
$$S \text{ is a } (C_K \delta)\text{-semiconvex set,}$$

that is, for any couple of points $x_0, x_1 \in S$ the ball centered at $x_{1/2} := (x_0 + x_1)/2$ of radius $C_K \delta |x_1 - x_0|^2$ intersects $S$. Since $S \subset B_K$, this implies that $\text{co}[S] \subset \mathcal{N}_{C_K' \delta}(S)$ for some positive constant $C_K'$ depending only on $K$. Thus, for any $\gamma \in (0, 1)$ we obtain

$$\mathcal{N}_{\delta^\gamma}(\text{co}[S]) \subset \mathcal{N}_{(1 + C_K')\delta^\gamma}(S).$$

Since $v = 1$ on $\partial\big(\mathcal{N}_{\delta^\gamma}(\text{co}[S])\big)$ and $\lambda_1/\lambda_2 \in (1/4, 4)$, by standard interior estimates for solution of the Monge-Ampère equation with constant right hand side (see for instance [28, Lemma 1.1]), we obtain

$$\underset{S}{\text{osc}}\, v \ \leq \ C_K'' \qquad\qquad\qquad\qquad\quad (6.61)$$

$$1 - C_K'' \delta^{\gamma/n} \ \leq \ v \leq 1 \qquad\qquad \text{on } \partial S, \qquad (6.62)$$

$$D^2 v \ \geq \ \delta^{\gamma/\tau}\, \text{Id}\,/C_K'' \qquad \text{in co}[S], \qquad (6.63)$$

for some $\tau > 0$ universal, and some constant $C_K''$ depending only on $K$.

Let us define

$$v^+ := (1 + 4\varepsilon + 2\sqrt{\delta})v - 4\varepsilon - 2\sqrt{\delta},$$

$$v^- := (1 - 4\varepsilon - \sqrt{\delta}/2)v + 4\varepsilon + \sqrt{\delta}/2 + 2C_K'' \delta^{\gamma/n}.$$

Our goal is to show that we can choose $\gamma$ universally small so that $v^- \geq u \geq v^+$ on $S$. Indeed, if we can do so, then by (6.61) this will imply (6.59), concluding the proof.

First of all notice that, thanks to (6.62), $v^- > u > v^+$ on $\partial S$. Let us show first that $v^+ \leq v$.

Assume by contradiction this is not the case. Then, since $u > v^+$ on $\partial S$,
$$\emptyset \neq Z := \{u < v^+\} \subset\subset S.$$
Since $v^+$ is convex, taking any supporting plane to $v^+$ at $x \in Z$, moving it down and then lifting it up until it touches $u$ from below, we deduce that
$$\nabla v^+(Z) \subset \nabla u(Z) \tag{6.64}$$
(recall that both $u$ and $v^+$ are of class $C^1$), thus by Remark 6.6
$$|\text{c-exp}(\nabla v^+(Z))| \leq |T_u(Z)|. \tag{6.65}$$
We show that this is impossible. For this, using (6.63) and choosing $\gamma := \tau/4$, for any $x \in Z$ we compute
$$\begin{aligned}
&D^2 v^+(x) + D_{xx}c\big(x, \text{c-exp}_x(\nabla v^+(x))\big) \\
&\geq (1 + \sqrt{\delta} + 4\varepsilon)D^2 v + \sqrt{\delta}D^2 v - \delta\,\text{Id} \\
&\geq (1 + \sqrt{\delta} + 4\varepsilon)D^2 v + (\delta^{1/4}/C_K'' - \delta)\,\text{Id} \\
&\geq (1 + \sqrt{\delta} + 4\varepsilon)D^2 v,
\end{aligned}$$
provided $\delta$ is sufficiently small, the smallness depending only on $K$. Thus, thanks to (6.58) we have
$$\begin{aligned}
&\frac{\det\big(D^2 v^+(x) + D_{xx}c(x, \text{c-exp}_x(\nabla v^+(x)))\big)}{\big|\det\big(D_{xy}c\big(x, \text{c-exp}_x(\nabla v^+(x))\big)\big)\big|} \\
&\geq \frac{\det\big((1 + \sqrt{\delta} + 4\varepsilon)D^2 v\big)}{1 + \delta} \\
&\geq (1 + \sqrt{\delta} + 4\varepsilon)^n (1 - 2\delta)\lambda_1/\lambda_2 \\
&\geq (1 + 4n\varepsilon)\lambda_1/\lambda_2.
\end{aligned}$$
In addition, thanks (6.63) and (6.58), since $\delta^{\gamma/\tau} = \delta^{1/4} \gg \delta$ we see that
$$D^2 v^+ > \|D_{xx}c\|_{C^0(B_K \times B_K)}\,\text{Id} \qquad \text{inside co}[S].$$
Hence, for any $x, z \in Z$, $x \neq z$ and $y = \text{c-exp}_x(\nabla v^+(x))$ (notice that $\text{c-exp}_x(\nabla v^+(x))$ is well-defined because of (6.64) and the assumption $\nabla u(S) \subset\subset \text{Dom c-exp}$), it follows
$$v^+(z) + c(z, y) \geq v^+(x) + c(x, y)$$
$$+ \frac{1}{2}\int_0^1 \Big(D^2 v^+\big(tz + (1-t)x\big) + D_{xx}c\big(tz + (1-t)x, y\big)\Big)(z - x) \cdot (z - x)\,dt$$
$$> v^+(x) + c(x, y),$$

where we used that $\nabla v^+(x) + D_x c(x, y) = 0$. This means that the supporting function $z \mapsto -c(z, y) + c(x, y) + v^+(x)$ can only touch $v^+$ from below at $x$, which implies that the map $Z \ni x \mapsto \text{c-exp}_x(\nabla v^+(x))$ is injective. Thus, by Lemma 6.9 we get

$$|\text{c-exp}(\nabla v^+(Z))| \geq (1 + 4n\varepsilon)\lambda_1/\lambda_2|Z|. \tag{6.66}$$

On the other hand, since $u$ is $C^1$, it follows from $(T_u)_\sharp f = g$ and (6.57) that

$$|T_u(Z)| = \int_Z \frac{f(x)}{g(T_u(x))} \, dx \leq \frac{\lambda_1(1 + \varepsilon)}{\lambda_2(1 - \varepsilon)} \leq (1 + 3\varepsilon)\lambda_1/\lambda_2|Z|.$$

This estimate combined with (6.66) shows that (6.65) is impossible unless $Z$ is empty. This proves that $v^+ \leq u$.

The proof of the inequality $v^- \leq u$ follows by the same argument expect for a minor modification. More precisely, let us assume by contradiction that $W := \{u > v^-\}$ is nonempty. In order to apply the previous argument we would need to know that $\nabla v^-(W) \subset \text{Dom c-exp}$. However, since the gradient of $v$ can be very large near $\partial S$, this may be a problem.

To circumvent this issue we argue as follows: since $W$ is nonempty, there exists a positive constant $\bar{\mu}$ such that $u$ touches $v^- + \bar{\mu}$ from below inside $S$. Let $E$ be the contact set, i.e., $E := \{u = v^- + \bar{\mu}\}$. Since both $u$ and $v^-$ are $C^1$, $\nabla u = \nabla v^-$ on $E$. Thus, if $\eta > 0$ is small enough, then the set $W_\eta := \{u > v^- + \bar{\mu} - \eta\}$ is nonempty and $\nabla v^-(W_\eta)$ is contained in a small neighborhood of $\nabla u(W_\eta)$, which is compactly contained in Dom c-exp. At this point, one argues exactly as in the first part of the proof, with $W_\eta$ in place of $Z$, to find a contradiction. □

**Theorem 6.11.** *Let $u, f, g, \eta_0, \delta_0$ be as in Theorem 6.5, and assume in addition that $c \in C^{k,\alpha}(B_3 \times B_3)$ and $f, g \in C^{k,\alpha}(B_{1/3})$ for some $k \geq 0$ and $\alpha \in (0, 1)$. There exist $\eta_1 \leq \eta_0$ and $\delta_1 \leq \delta_0$ small, such that, if*

$$\|f - \mathbf{1}_{C_1}\|_\infty + \|g - \mathbf{1}_{C_2}\|_\infty \leq \delta_1, \tag{6.67}$$

$$\|c(x, y) + x \cdot y\|_{C^2(B_3 \times B_3)} \leq \delta_1, \tag{6.68}$$

*and*

$$\left\|u - \frac{1}{2}|x|^2\right\|_{C^0(B_3)} \leq \eta_1, \tag{6.69}$$

*then $u \in C^{k+2,\alpha}(B_{1/9})$.*

*Proof.* We divide the proof in two steps.

- *Step* 1 : $C^{1,1}$ *regularity.* Fix a point $x_0 \in B_{1/8}$, $y_0 = $ c-exp$_{x_0}(\nabla u(x_0))$. Up to replace $u$ (resp. $c$) with the function $u_1$ (resp. $c_1$) constructed in Steps 4 and 5 in the proof of Theorem 6.5, we can assume that $u \geq 0$, $u(0) = 0$, that

$$S_h := S(0, 0, u, h) = \{u \leq h\},$$

and that

$$D_{xy}c(0, 0) = -\text{Id}. \tag{6.70}$$

Under these assumptions we will show that the sections of $u$ are of "good shape", *i.e.*,

$$B_{\sqrt{h}/K} \subset S_h \subset B_{K\sqrt{h}} \qquad \forall h \leq h_1, \tag{6.71}$$

for some universal $h_1$ and $K$. Arguing as in Step 6 of Theorem 6.5, this will give that $u$ is $C^{1,1}$ at the origin, and thus at every point in $B_{1/8}$.

First of all notice that, thanks to (6.69), for any $h_1 > 0$ we can choose $\eta_1 = \eta_1(h_1) > 0$ small enough such that (6.71) holds for $S_{h_1}$ with $K = 2$. Hence, assuming without loss of generality that $\delta_1 \leq 1$, we see that

$$B_{\sqrt{h_1}/3} \subset \mathcal{N}_{\delta_1^\gamma \sqrt{h_1}}(\text{co}[S_{h_1}]) \subset B_{3\sqrt{h_1}},$$

where $\gamma$ is the exponent from Lemma 6.10. Let $v_1$ solve the Monge-Ampère equation

$$\begin{cases} \det(D^2 v_1) = f(0)/g(0) & \text{in } \mathcal{N}_{\delta_1^\gamma \sqrt{h_1}}(\text{co}[S_{h_1}]), \\ v_1 = h_1 & \text{on } \partial \mathcal{N}_{\delta_1^\gamma \sqrt{h_1}}(\text{co}[S_{h_1}]). \end{cases}$$

Since $B_{1/3} \subset N_{\delta_1^\gamma \sqrt{h_1}}(\text{co}[S_{h_1}])/\sqrt{h_1} \subset B_3$, by standard Pogorelov estimates applied to the function $v_1(\sqrt{h_1}x)/h_1$ (see Section 2.3), it follows that $|D^2 v_1(0)| \leq M$, with $M > 0$ some large universal constant.

Let $h_k := h_1 2^{-k}$ and define $\bar{K} \geq 3$ to be the largest number such that any solution $w$ of

$$\begin{cases} \det(D^2 w) = f(0)/g(0) & \text{in } Z, \\ w = 1 & \text{on } \partial Z, \end{cases} \tag{6.72}$$

with $B_{1/\bar{K}} \subset Z \subset B_{\bar{K}}$ satisfies $|D^2 w(0)| \leq M + 1$ [7]. We prove by

---

[7] The fact that $\bar{K}$ is well defined (*i.e.*, $3 \leq \bar{K} < \infty$) follows by the following facts: first of all, by definition, $M$ is an a-priori bound for $|D^2 w(0)|$ whenever $w$ is a solution of (6.72) with $B_{1/3} \subset Z \subset B_3$, so $\bar{K} \geq 3$. On the other hand $\bar{K} \leq \sqrt{M+1}$, since the function

$$\bar{w} := (M+1)x_1^2 + \frac{x_2^2}{M+1} + x_3^2 + \ldots + x_n^2$$

is a solution of (6.72) with $B_{1/\sqrt{M+1}} \subset Z \subset B_{\sqrt{M+1}}$ and $|D^2 \bar{w}(0)| = 2(M+1)$, see also Theorem 2.16 and the discussion below it.

induction that (6.71) holds with $K = \bar{K}$.

If $h = h_1$ then we already know that (6.71) holds with $K = 2$ (and so with $K = \bar{K}$).

Assume now that (6.71) holds with $h = h_k$ and $K = \bar{K}$, and we want to show that it holds with $h = h_{k+1}$. For this, for any $k \in \mathbb{N}$ we consider $u_k$ the solution of

$$\begin{cases} \det(D^2 v_k) = f(0)/g(0) & \text{in } \mathcal{N}_{\delta_k^\gamma \sqrt{h_k}}(\mathrm{co}[S_{h_k}]), \\ v_k = h_1 2^{-k} & \text{on } \partial \mathcal{N}_{\delta_k^\gamma \sqrt{h_k}}(\mathrm{co}[S_{h_k}]), \end{cases}$$

where

$$\delta_k := \|c(x, y) + x \cdot y\|_{C^2(S_{h_k} \times T_u(S_{h_k}))} \leq \delta_1.$$

Let us consider the rescaled functions

$$\bar{u}_k(x) := u\big(\sqrt{h_k}x\big)/h_k \quad \bar{v}_k(x) := v_k\big(\sqrt{h_k}x\big)/h_k.$$

Since by the inductive hypothesis $B_{1/\bar{K}} \subset \bar{S}_k := \{\bar{u}_k \leq 1\} \subset B_{\bar{K}}$, we can apply Lemma 6.10 to deduce that

$$\begin{aligned} \|\bar{u}_k - \bar{v}_k\|_{C^0(\bar{S}_k)} &\leq C_{\bar{K}}\Big( \underset{S_{h_k}}{\mathrm{osc}}\, f + \underset{T_u(S_{h_k})}{\mathrm{osc}}\, g + \delta_k^{\gamma/n} \Big) \\ &\leq C_{\bar{K}}(\delta_1 + \delta_1^{\gamma/n}). \end{aligned} \tag{6.73}$$

This implies in particular that, if $\delta_1$ is sufficiently small, $B_{1/(2\bar{K})} \subset \{\bar{v}_k \leq 1\} \subset B_{2\bar{K}}$. By Proposition 2.12 (ii) and Remark 2.14, the shapes of $\{\bar{v}_k \leq 1\}$ and $\{\bar{v}_k \leq 1/2\}$ are comparable, moreover they are well included into each other: there exists a universal constant $L > 1$ such that

$$B_{1/(L\bar{K})} \subset \{\bar{v}_k \leq 1/2\} \subset B_{L\bar{K}}, \quad \mathrm{dist}\big(\{\bar{v}_k \leq 1/4\}, \partial\{\bar{v}_k \leq 1/2\}\big) \geq 1/(LK).$$

Using again (6.73) we deduce that, if $\delta_1$ is sufficiently small,

$$B_{1/(2L\bar{K})} \subset \{\bar{u}_k \leq 1/2\} \subset B_{2L\bar{K}}, \quad \mathrm{dist}\big(\{\bar{u}_k \leq 1/4\}, \partial\{\bar{u}_k \leq 1/2\}\big) \geq 1/(2LK)$$

that is, scaling back,

$$B_{\sqrt{h_{k+1}}/(2L\bar{K})} \subset S_{h_{k+1}} \subset B_{2L\bar{K}\sqrt{h_{k+1}}}, \quad \mathrm{dist}\big(S_{h_{k+2}}, \partial S_{h_{k+1}}\big) \geq \sqrt{h_k}/(2LK) \tag{6.74}$$

This allows us to apply Lemma 6.10 also to $\bar{u}_{k+1}$ to get

$$\|\bar{u}_{k+1} - \bar{v}_{k+1}\|_{C^0(\bar{S}_{k+1})} \leq C_{2L\bar{K}}\Big( \underset{S_{h_{k+1}}}{\mathrm{osc}}\, f + \underset{T_u(S_{h_{k+1}})}{\mathrm{osc}}\, g + \delta_{k+1}^{\gamma/n} \Big). \tag{6.75}$$

We now observe that, by (6.71) and the $C^{1,\beta}$ regularity of $u$ (Theorem 6.5) it follows that

$$\text{diam}(S_{h_k}) + \text{diam}(T_u(S_{h_k})) \leq C h_k^{\beta/2},$$

so by the $C^{0,\alpha}$ regularity of $f$ and $g$, and the $C^{2,\alpha}$ regularity of $c$, we have (recall that $\gamma < 1$)

$$\underset{S_{h_k}}{\text{osc}}\, f + \underset{T_u(S_{h_k})}{\text{osc}}\, g + \delta_k^{\gamma/n} \leq C' h_k^\sigma, \qquad \sigma := \frac{\alpha\beta\gamma}{2n} \qquad (6.76)$$

Hence, by (6.73) and (6.75),

$$\|\bar{u}_k - \bar{v}_k\|_{C^0(\bar{S}_k)} + \|\bar{u}_{k+1} - \bar{v}_{k+1}\|_{C^0(\bar{S}_{k+1})} \leq C \left( C_{\bar{K}} + C_{2L\bar{K}} \right) h_k^\sigma,$$

from which we deduce (recall that $h_k = 2h_{k+1}$)

$$\begin{aligned}
\|v_k - v_{k+1}\|_{C^0(S_{h_{k+1}})} &\leq \|v_k - u\|_{C^0(S_{h_k})} + \|u - v_{k+1}\|_{C^0(S_{h_{k+1}})} \\
&= h_k \|\bar{u}_k - \bar{v}_k\|_{C^0(S_k)} + h_{k+1} \|\bar{u}_{k+1} - \bar{v}_{k+1}\|_{C^0(S_{k+1})} \\
&\leq C \left( C_{\bar{K}} + C_{2L\bar{K}} \right) h_k^{1+\sigma}.
\end{aligned}$$

Since $v_k$ and $v_{k+1}$ are two strictly convex solutions of the Monge Ampère equation with constant right hand side inside $S_{h_{k+1}}$, and since $S_{h_{k+2}}$ is "well contained" inside $S_{h_{k+1}}$, by classical Pogorelov and Schauder estimates we get

$$\|D^2 v_k - D^2 v_{k+1}\|_{C^0(S_{h_{k+2}})} \leq C'_{\bar{K}} h_k^\sigma \qquad (6.77)$$

$$\|D^3 v_k - D^3 v_{k+1}\|_{C^0(S_{h_{k+2}})} \leq C'_{\bar{K}} h_k^{\sigma-1/2}, \qquad (6.78)$$

where $C'_{\bar{K}}$ is some constant depending only on $\bar{K}$. By (6.77) applied to $v_j$ for all $j = 1, \ldots, k$ (this can be done since, by the inductive assumption, (6.71) holds for $h = h_j$ with $j = 1, \ldots, k$) we obtain

$$\begin{aligned}
|D^2 v_{k+1}(0)| &\leq |D^2 v_1(0)| + \sum_{j=1}^k |D^2 v_j(0) - D^2 v_{j+1}(0)| \\
&\leq M + C'_{\bar{K}} h_1^\sigma \sum_{j=0}^k 2^{-j\sigma} \\
&\leq M + \frac{C'_{\bar{K}}}{1 - 2^{-\sigma}} h_1^\sigma \leq M + 1,
\end{aligned}$$

provided we choose $h_1$ small enough (recall that $h_k = h_1 2^{-k}$). By the definition of $\bar{K}$ it follows that also $S_{h_{k+1}}$ satisfies (6.71), concluding the proof of the inductive step.

• *Step* 2 : *higher regularity.* Now that we know that $u \in C^{1,1}(B_{1/8})$, Equation (1.38) becomes uniformly elliptic. So one may use Evans-Krylov Theorem (see [25]) to obtain that $u \in C^{2,\sigma'}_{\text{loc}}(B_{1/9})$ for some $\sigma' > 0$, and then standard Schauder estimates to conclude the proof. However, for convenience of the reader, we show here how to give a simple direct proof of the $C^{2,\sigma'}$ regularity of $u$ with $\sigma' = 2\sigma$.

As in the previous step, it suffices to show that $u$ is $C^{2,\sigma'}$ at the origin, and for this we have to prove that there exists a sequence of paraboloids $P_k$ such that

$$\sup_{B_{r_0^k/C}} |u - P_k| \leq C r_0^{k(2+\sigma')} \tag{6.79}$$

for some $r_0, C > 0$.

Let $v_k$ be as in the previous step, and let $P_k$ be their second order Taylor expansion at $\mathbf{0}$:

$$P_k(x) = v_k(\mathbf{0}) + \nabla v_k(\mathbf{0}) \cdot x + \frac{1}{2} D^2 v_k(\mathbf{0}) x \cdot x.$$

We observe that, thanks to (6.71),

$$
\begin{aligned}
\|v_k - P_k\|_{C^0(B(0,\sqrt{h_{k+2}/C}))} &\leq \|v_k - P_k\|_{C^0(S_{h_{k+2}})} \\
&\leq C\|D^3 v_k\|_{C^0(S_{h_{k+2}})} h_k^{3/2}.
\end{aligned}
\tag{6.80}
$$

In addition, by (6.78) applied with $j = 1, \ldots, k$ and recalling that $h_k = h_1 2^{-k}$ and $2\sigma < 1$ (see (6.76)), we get

$$
\begin{aligned}
\|D^3 v_k\|_{C^0(S_{h_{k+2}})} &\leq \|D^3 v_1\|_{C^0(S_{h_3})} + \sum_{j=1}^{k} \|D^3 v_j - D^3 v_{j+1}\|_{C^0(S_{h_{j+2}})} \\
&\leq C\left(1 + \sum_{j=1}^{k} h_j^{(\sigma-1/2)}\right) \leq C h_k^{\sigma-1/2}.
\end{aligned}
\tag{6.81}
$$

Combining (6.71), (6.80), (6.81), and recalling (6.73) and (6.76), we obtain

$$
\begin{aligned}
\|u - P_k\|_{C^0(B_{\sqrt{h_{k+2}}/K})} &\leq \|v_k - P_k\|_{C^0(S_{h_{k+2}})} + \|v_k - u\|_{C^0(S_{h_{k+2}})} \\
&\leq C h_k^{1+\sigma},
\end{aligned}
$$

so (6.79) follows with $r_0 = 1/\sqrt{2}$ and $\sigma' = 2\sigma$. □

# Appendix A
# Properties of convex functions

In this appendix we report the main properties of convex (and semiconvex) functions which we have used in the previous Chapters.

A function $u : \mathbb{R}^n \to [-\infty, +\infty]$ is said convex if its epigraph:

$$\text{Epi}(u) := \{(x, t) \in \mathbb{R}^n \times \mathbb{R} : u(x) \leq t\}$$

is a convex subset of $\mathbb{R}^{n+1}$. In case $u > -\infty$ the above definition is equivalent to ask that

$$u(tx + (1-t)y) \leq tu(x) + (1-t)u(y), \qquad \forall x, y \in \mathbb{R}^n \ t \in [0, 1].$$

A convex function is said *lower semicontinuous* if $\text{Epi}(u)$ is closed and *proper* if $u(x_0) > -\infty$ for some $x_0$. It is easy to see that for a proper and lower semicontinuous function $\{u = -\infty\}$ is empty. We define the domain of $u$ as the convex set

$$\text{Dom}(u) = \{u < +\infty\}$$

and in the sequel we will always assume that $\text{Dom}(u)$ has non-empty interior and that $u$ is proper and lower semicontinuous.

An important role in convex analysis is played by the subdifferential of a convex function. A point $x$ is in the domain of the subdifferential, $x \in \text{Dom}(\partial u)$, if there exists a non-vertical supporting plane to $\text{Epi}(u)$ at the point $(x, u(x))$. By classical theorems, see [85, Chapter 12], it is easy to see that

$$\text{Int}(\text{Dom}(u)) \subset \text{Dom}(\partial u) \subset \text{Dom}(u).$$

If $x \in \text{Dom}(\partial u)$ the subdifferential of $x$ at $u$ is defined as the set of the slopes of the supporting planes:

$$\partial u(x) = \{p \in \mathbb{R}^n : u(y) \geq u(x) + p \cdot (y - x)\}.$$

In particular $p \in \partial u(x)$ if and only if the function

$$y \mapsto u(y) - p \cdot y$$

has a minimum at $x$. Since the function $u(y) - p \cdot y$ is clearly convex, any local minimum is a global minimum, thus the above observation leads to the following useful characterization:

$$\partial u(x) = \{p \in \mathbb{R}^n : u(y) \geq u(x) + p \cdot (y-x) \text{ for all } y \text{ in a neighborhood of } x\}.$$

It is immediate to see that $\partial u(x)$ is a convex subset of $\mathbb{R}^n$. Moreover, as subset of $\mathbb{R}^n \times \mathbb{R}^n$, the graph of the subdifferential is closed:

$$x_k \to x, \quad \partial u(x_k) \ni p_k \to p \quad \Rightarrow \quad x \in \text{Dom}(\partial u), \quad p \in \partial u(x).$$

We also recall the following elementary estimate: if $\Omega' \Subset \Omega'' \subset \text{Dom}(u)$ then

$$\sup_{x \in \Omega'} \sup_{p \in \partial u(x)} |p| \leq \frac{\text{osc}_{\Omega''} u}{\text{dist}(\Omega', \partial \Omega)}. \tag{A.1}$$

Since convex functions are locally bounded on their domains (see [85, Theorem 10.1]), thanks to the "non smooth" mean value theorem (Lemma A.1 below) we see that convex functions are also locally Lipschitz.

Given a proper and l.s.c. function $u$ its *convex conjugate* (or Legendre transform) is the convex and lower semicontinuous function:

$$u^*(p) = \sup_{x \in \mathbb{R}^n} \{p \cdot x - u(x)\}.$$

Since $u$ is proper and lower semicontinuous one can verify that

$$u^{**}(x) = \sup_{p \in \mathbb{R}^n} \{p \cdot x - u^*(p)\} = u(x).$$

In addition

$$p \in \partial u(x) \iff x \in \partial u^*(p),$$

and in this case

$$u(x) = p \cdot x - u^*(p) \quad \text{and} \quad u^*(p) = p \cdot x - u(x).$$

Indeed

$$p \in \partial u(x) \iff u(y) - p \cdot y \geq u(x) - p \cdot x \quad \forall y$$
$$\iff u^*(p) = p \cdot x - u(x)$$
$$\iff u^*(p) - p \cdot x \leq u^*(q) - q \cdot x \quad \forall q$$
$$\iff x \in \partial u^*(p).$$

In particular $\partial u$ and $\partial u^*$ are (as multivalued maps) one the inverse of the other.

It is immediate to verify that if $u$ is differentiable at $x$ then $\partial u(x) = \{\nabla u(x)\}$. To show the converse implication we need the following.

**Lemma A.1 (Non smooth mean value theorem).** *Let $u$ be a convex and finite function, if $x, y \in \mathrm{Int}(\mathrm{Dom}(u))$ then there exist $z \in (x, y)$ [1] and $p \in \partial u(z)$ such that*

$$u(y) - u(x) = p \cdot (y - x)$$

*Proof.* Let $\Omega \Subset \mathrm{Dom}(u)$ be a convex set such that $[x, y] \subset \Omega$ and let us consider the regularized functions

$$u_\varepsilon(w) = \int u(y)\varphi_\varepsilon(w - y)dy$$

with $\varphi_\varepsilon$ a family of compactly supported mollifiers. Then $u_\varepsilon$ are defined and convex on a $\varepsilon$ neighborhood of $\Omega$ and $u_\varepsilon$ uniformly converge to $u$ in $\Omega$ (recall that any pointwise bounded sequence of convex functions is locally bounded, see [85, Theorem 10.6]), hence, since the $u_\varepsilon$ are smooth, (A.1) implies that $u_\varepsilon$ are locally Lipschitz). By the classical mean value theorem for smooth functions there exists a point $z_\varepsilon \in (x, y)$ such that

$$u_\varepsilon(x) - u_\varepsilon(y) = \nabla u_\varepsilon(z_\varepsilon) \cdot (y - x).$$

Passing to the limit as $\varepsilon$ goes to $0$ we see that, up to subsequence, $z_\varepsilon \to z \in [x, y]$ and $\nabla u(z_\varepsilon) \to p \in \partial u(z)$. Hence

$$u(y) - u(x) = p \cdot (y - x), \qquad p \in \partial u(z), \; z \in [x, y]. \qquad (A.2)$$

If $z \in (x, y)$ we are done. In case $z = x$, for instance, by the above equality it is immediate to see that $p \in \partial u(w)$ for all $w \in [x, y]$ and thus (A.2) holds also for some $\bar{z} \in (x, y)$, proving the lemma. $\qquad \square$

**Lemma A.2.** *Assume that $x \in \mathrm{Int}(\mathrm{Dom}(u))$ and that $\partial u(x) = \{p\}$, then $u$ is differentiable at $x$ and $p = \nabla u(x)$.*

*Proof.* We want to show that

$$u(y) = u(x) + p \cdot (y - x) + o(|y - x|).$$

By Lemma A.1, there exists a point $z \in (x, y)$ such that

$$u(y) = u(x) + q_z \cdot (y - x) \quad q_z \in \partial u(z).$$

If $y \to x$ then also $z \to x$ and, by (A.1) and the closure of the subdifferential, $q_z \to \bar{q} \in \partial u(x)$. Since, by assumption, $\partial u(x)$ is a singleton, $q_z \to p$. Then

$$u(y) = u(x) + p \cdot (y - x) + (q_z - p) \cdot (y - x)$$
$$= u(x) + p \cdot (y - x) + o(|y - x|). \qquad \square$$

---

[1] $(x, y)$ denotes the open segment with extremes $x$ and $y$ while $[x, y]$ is the closed segment with the same extremes.

We have already shown that convex functions are locally Lipschitz, hence, by Rademacher Theorem [47], they are differentiable almost everywhere on their domain. Actually, thanks to the above lemma, one can give a more elementary proof of this fact using the following ingredients:

- The restriction of a convex function to a line is a one-dimensional convex function.
- One-dimensional convex functions are differentiable outside a countable set (this follows by the monotonicty of the incremental ratio).
- By Fubini Theorem a convex function admits partial derivative almost everywhere.
- If in a point a convex function is derivable along $n$ independent directions, then $\partial u(x)$ is a singleton.

By the above considerations, for every $x \in \text{Int}(\text{Dom}(u))$ the set of *reachable gradients*

$$\nabla_* u(x) = \Big\{ p \in \partial u(x) : \text{exists a sequence of differentiability points}$$
$$x_k, x_k \to x, \nabla u(x_k) \to p \Big\}$$

is non empty. Actually they are enough to generate the full subdifferential:

**Proposition A.3.** *If $u$ is convex and $x \in \text{Int}(\text{Dom}(u))$ then*

$$\partial u(x) = \overline{\text{co}\left[\nabla_* u(x)\right]}.$$

*Proof.* Let

$$C = \overline{\text{co}\left[\nabla_* u(x)\right]}.$$

Being $\partial u(x)$ closed and convex $C \subset \partial u(x)$. Let us assume that there exists $\bar{p} \in \partial u(x) \setminus C$. Since $C$ is a compact and convex (recall that, by (A.1), $\partial u(x)$ is bounded), there exists a vector $e \in \mathbb{S}^{n-1}$ such that

$$e \cdot \bar{p} := 4\delta > 0 \geq e \cdot p \quad \forall p \in C.$$

Since $C$ is compact there exists a small $\sigma$ such that for all $v \in \overline{B}_\sigma(e)$

$$v \cdot \bar{p} \geq 3\delta > \delta \geq v \cdot p \quad \forall p \in C. \tag{A.3}$$

Since the cone generated by $\overline{B}_\sigma(e)$ and the origin has positive measure we can find a sequence $v_k \in \overline{B}_\sigma(e)$ and $t_k \to 0$ such that $x + t_k v_k \to x$ and

$u$ is differentiable at $x + t_k v_k$. Bu the monotonicity of the subdifferential (recall Theorem 1.6)

$$t(\nabla u(x_k + t v_k) - \bar{p}) \cdot v_k \geq 0.$$

Up to a subsequence, $v_k \to \bar{v} \in \overline{B}_\sigma(e)$ and $\nabla u(x_k + t v_k) \to q \in \nabla_* u(x) \subset C$. Then

$$q \cdot \bar{v} \geq \bar{p} \cdot \bar{v},$$

contarddicting (A.3).  □

We conclude this appendix investigating second order properties of convex functions. The first observation is that distributional second order derivatives of convex functions are measures (*i.e.* distributions of order zero). In particular the almost everywhere defined function $\nabla u(x)$ belongs to $BV_{\text{loc}}$, see [8] for main properties of $BV$ functions.
The heuristic idea behind the proof is that for a (smooth) convex function $D^2 u \geq 0$ and that any positive distribution is a measure.

**Proposition A.4.** *Let $\Omega \Subset \text{Dom}(u)$ then $D^2 u$ is a symmetric matrix valued Radon measure in $\Omega$.*

*Proof.* Let $\Omega \Subset \Omega' \Subset \text{Dom}(u)$ and $u_\varepsilon$ be a sequence of smooth convex functions defined on $\Omega''$ uniformly converging to $u$ in $\Omega'$ (for instance we can take as $u_\varepsilon$ the convolution of $u$ with a family of mollifiers as introduced in the proof of Lemma A.1). Clearly

$$D^2 u_\varepsilon \to D^2 u$$

as distributions. By Riesz representation theorem, it is enough to show that

$$\limsup_{\varepsilon \to 0} \int_\Omega |D^2 u_\varepsilon| \leq C.$$

for some constant $C$ independent on $\varepsilon$. To see this recall that, by convexity, $|D^2 u_\varepsilon| \leq n \Delta u_\varepsilon$, hence (assuming without loss of generality that $\Omega$ is smooth)

$$\int_\Omega |D^2 u_\varepsilon| \leq n \int_\Omega \Delta u_\varepsilon = n \int_{\partial \Omega} \nabla u_\varepsilon \cdot \nu_{\partial \Omega}$$

$$\leq n \mathcal{H}^{n-1}(\partial \Omega) \sup_\Omega |\nabla u_\varepsilon| \leq n \mathcal{H}^{n-1}(\partial \Omega) \frac{\sup_{\Omega'} |u_\varepsilon|}{\text{dist}(\Omega, \partial \Omega')}.$$

Since $u_\varepsilon$ converge uniformly to $u$ on $\Omega'$

$$\limsup_{\varepsilon \to 0} \int_\Omega |D^2 u_\varepsilon| \leq n \mathcal{H}^{n-1}(\partial \Omega) \frac{\sup_{\Omega'} |u|}{\text{dist}(\Omega, \partial \Omega')},$$

concluding the proof.  □

By the Lebesgue- Radon-Nykodim Theorem $D^2 u$ can be decomposed as

$$D^2 u = \nabla^2 u \, d\mathscr{L} + (D^2 u)^s.$$

where $\nabla^2 u$ is defined as

$$\nabla^2 u(x) = \lim_{r \to 0} \frac{D^2 u(B_r(x))}{|B_r(x)|} \quad (^2).$$

A well know property of $BV$ functions is that they are approximately differentiable almost everywhere[3] (see [8, Theorem 3.83]). In particular $\nabla u(x)$ is approximately differentiable with approximate differential equal to $\nabla^2 u$. The following theorem, due to Aleksandrov, implies that actually $\nabla u$ is differentiable almost everywhere.

**Theorem A.5 (Aleksandrov).** *Let $u$ be a convex function, then for almost every point of differentiability $x$ in the interior of its domain the following equivalent properties hold true:*

(i) $u(y) = u(x) + \nabla u(x) \cdot (y-x) + \frac{1}{2} \nabla^2 u(x)(y-x) \cdot (y-x) + o(|y-x|^2)$
(ii) $\partial u(y) = \nabla u(x) + \nabla^2 u(x) \cdot (y - x) + o(|y - x|).$

*Here in* (ii) *we mean that*

$$p = \nabla u(x) + \nabla^2 u(x) \cdot (y - x) + o(|y - x|) \quad \text{for all } p \in \partial u(y)$$

*Proof.* We first show the equivalence of (i) and (ii). To see that (ii) implies (i) one simply applies Lemma A.1. Let us show that (i) implies (ii). For this assume, by sake of contradiction, that there exists a sequence of points $y_k$ converging to $x$ such that for some $p_k \in \partial u(y_k)$

$$\liminf_{k \to 0} \frac{|p_k - \nabla u(x) - \nabla^2 u(x)(y_k - x)|}{|y_k - x|} > 0. \qquad (A.4)$$

We can clearly assume that $x = 0 = |\nabla u(x)|$, let us write $y_k = t_k v_k$ with $v_k \in \mathbb{S}^{n-1}$ and consider the convex functions

$$u_k(w) = \frac{u(t_k w)}{t_k^2} \quad w \in B_2.$$

---

[2] It is part of the theorem that the above limit exists almost everywhere. Notice that in particular this implies that $\nabla^2 u$ is symmetric.

[3] A function $v$ is approximate differentiable at $x$ if there exists a linear function $L_x$ such that

$$\frac{v(x + rh) - v(x)}{r} \to L_x h$$

locally in measure. In this case $L_x$ is unique and it is called the approximate differential of $v$ at $x$

By our assumptions $u_k \to \bar{u} := \nabla^2 u(0)w \cdot w/2$ locally uniformly in $B_2$. Up to subsequence $v_k \to v$. Moreover $p_k/t_k \in \partial u_k(v_k)$. Since $u_k$ are locally bounded by (A.1), $p_k/t_k$ is bounded, hence, always up to subsequence,

$$\partial u_k(v_k) \ni p_k/t_k \to p \in \partial \bar{u}(v) = \{\nabla^2 u(0)v\},$$

contradicting (A.4).

We now show that (i) holds almost everywhere. For this let $x$ be a point such that

$$\frac{|(D^2 u)^s|(B_r(x))}{|B_r(x)|} \to 0 \quad \text{and} \quad \lim_{r \to 0} \fint_{B_r(x)} |\nabla^2 u(y) - \nabla^2 u(x)| dy \to 0$$

(A.5)

as $r \to 0$. Notice that the above properties hold true for almost every $x$. Let us assume that $x = 0 = |\nabla u(x)|$, our claim is equivalent to show that

$$\lim_{t \to 0} \frac{u(tv)}{t^2} = \frac{1}{2}\nabla^2 u(0)v \cdot v \quad \text{uniformly in } B_1.$$

Since the functions $u(tv)/t^2$ are convex, $L^1$ convergence implies local uniform convergence, so it is enough to show

$$\lim_{t \to 0} \int_{B_2} \left| \frac{u(tv)}{t^2} - \frac{1}{2}\nabla^2 u(0)v \cdot v \right| dv = 0.$$

By a change of variable the above is implied by

$$\lim_{r \to 0} \fint_{B_r} \left| \frac{u(x) - \frac{1}{2}\nabla^2 u(0)x \cdot x}{r^2} \right| dx = 0.$$

To see that the above limit is zero let us introduce the regularization of $u$

$$u_\varepsilon(x) = \int u(y)\varphi_\varepsilon(x - y) dy,$$

where

$$\varphi_\varepsilon(z) = \frac{1}{\varepsilon^n}\varphi\left(\frac{z}{\varepsilon}\right)$$

for some smooth and positive probability density $\varphi$ supported in $B_1$. Since $u_\varepsilon(x) \to u(x)$ and $\nabla u_\varepsilon(0) \to \nabla u(0) = 0$ by Fatou Lemma:

$$\limsup_{r \to 0} \fint_{B_r} \left| \frac{u(x) - \frac{1}{2}\nabla^2 u(0)x \cdot x}{r^2} \right| dx$$

$$\leq \limsup_{r \to 0} \liminf_{\varepsilon \to 0} \fint_{B_r} \left| \frac{u_\varepsilon(x) - \nabla u_\varepsilon(0) \cdot x - \frac{1}{2}\nabla^2 u(0)x \cdot x}{r^2} \right| dx. \quad \text{(A.6)}$$

By Taylor formula, Fubini Theorem, and a change of variable

$$\limsup_{r\to 0}\liminf_{\varepsilon\to 0}\fint_{B_r}\left|\frac{u_\varepsilon(x)-\nabla u_\varepsilon(0)\cdot x-\frac{1}{2}\nabla^2 u(0)x\cdot x}{r^2}\right|dx$$

$$\leq \limsup_{r\to 0}\liminf_{\varepsilon\to 0}\fint_{B_r}\int_0^1\left|\frac{D^2 u_\varepsilon(sx)x\cdot x-\nabla^2 u(0)x\cdot x}{r^2}\right|(1-s)dsdx$$

$$= \limsup_{r\to 0}\liminf_{\varepsilon\to 0}\int_0^1(1-s)\fint_{B_r}\left|\frac{D^2 u_\varepsilon(sx)x\cdot x-\nabla^2 u(0)x\cdot x}{r^2}\right|dxds$$

$$= \limsup_{r\to 0}\liminf_{\varepsilon\to 0}\int_0^1(1-s)\fint_{B_{sr}}\left|\frac{D^2 u_\varepsilon(y)y\cdot y-\nabla^2 u(0)y\cdot y}{(sr)^2}\right|dyds$$

$$\leq \limsup_{r\to 0}\liminf_{\varepsilon\to 0}\int_0^1\fint_{B_{rs}}|D^2 u_\varepsilon(y)-\nabla u(0)|dyds.$$

$$(A.7)$$

Now, for every radius $\rho$, by Fubini Theorem and the definition of $\varphi_\varepsilon$,

$$\int_{B_\rho}|D^2 u_\varepsilon(y)-\nabla^2 u(0)|dy$$

$$=\int_{B_\rho}\left|\int \varphi_\varepsilon(x-y)d(D^2 u)^s(x)+\int \varphi_\varepsilon(x-y)\big(\nabla^2 u(x)-\nabla^2 u(0)\big)dx\right|dy$$

$$\leq \int_{B_{\rho+\varepsilon}}\left(\int_{B_\rho}\varphi_\varepsilon(x-y)dy\right)d|(D^2 u)^s|(x)$$

$$+\int_{B_{\rho+\varepsilon}}|\nabla^2 u(x)-\nabla^2 u(0)|\left(\int_{B_\rho}\varphi_\varepsilon(x-y)dy\right)dx$$

$$\leq \sup_{w\in B_{\rho+\varepsilon}}\frac{|B_\rho\cap B_\varepsilon(w)|}{\varepsilon^n}\left(|(D^2 u)^s|(B_{\rho+\varepsilon})+\int_{B_{\rho+\varepsilon}}|\nabla^2 u(x)-\nabla^2 u(0)|dx\right)$$

$$\leq C(n)\frac{\min\{\varepsilon^n,\rho^n\}}{\varepsilon^n}\left(|(D^2 u)^s|(B_{\rho+\varepsilon})+\int_{B_{\rho+\varepsilon}}|\nabla^2 u(x)-\nabla^2 u(0)|dx\right),$$

hence,

$$\fint_{B_\rho}|D^2 u_\varepsilon(y)-\nabla^2 u(0)|dy$$

$$\leq C(n)\frac{\min\{\varepsilon^n,\rho^n\}(\rho+\varepsilon)^n}{(\rho\varepsilon)^n}\left(\frac{|(D^2 u)^s|(B_{\rho+\varepsilon})}{|B_{\rho+\varepsilon}|}+\fint_{B_{\rho+\varepsilon}}|\nabla^2 u(x)-\nabla^2 u(0)|dx\right)$$

$$\leq C(n)\left(\frac{|(D^2 u)^s|(B_{\rho+\varepsilon})}{|B_{\rho+\varepsilon}|}+\fint_{B_{\rho+\varepsilon}}|\nabla^2 u(x)-\nabla^2 u(0)|dx\right).$$

which, by (A.5), is uniformly bounded for $\varepsilon$ and $\rho$ small. Hence (as a function of $s$) the integrand in the right hand side of (A.7) is uniformly bounded. Moreover, by (A.5) and the above equation

$$\limsup_{r \to 0} \limsup_{\varepsilon \to 0} \fint_{B_{sr}} |D^2 u_\varepsilon(y) - \nabla^2 u(0)| dy = 0.$$

Recalling (A.6) and applying Dominated Convergence Theorem to the right hand side of (A.7), we finally conclude the proof.  □

A function $u$ is said $C$-semiconvex if $u - C|x|^2/2$ is convex or, equivalently, if $D^2 u \geq C$ Id in the sense of distributions, notice that the sum of a $C$-semiconvex function $u$ and of a $C^2$ function $v$ is $C + \|v\|_{C^2}$-semiconvex. It is clear that, since a semiconvex function is a smooth perturbation of a convex one, all the above properties of convex functions still hold true for semiconvex functions, in this case the role of the subdifferential is played by the Frechet subdifferential

$$\partial^- u(x) = \left\{ p \in \mathbb{R}^n : \quad u(y) \geq u(x) + p \cdot (y - x) + o(|y - x|) \right\}. \quad \text{(A.8)}$$

We conclude the appendix with the following useful interpolation inequality which is the equivalent of (A.1) for semiconvex functions.

**Lemma A.6.** *Let $u : \Omega \to \mathbb{R}$ be a $C$-semiconvex function and let $K \subset \Omega$ be a compact set, then*

$$\sup_{x \in K} \sup_{p \in \partial^- u(x)} |p| \leq \frac{\mathrm{osc}_\Omega\, u}{\mathrm{dist}(K, \partial\Omega)} + \sqrt{C_-\, \mathrm{osc}_\Omega\, u}, \quad \text{(A.9)}$$

*where $C_- = \max\{-C, 0\}$.*

*Proof.* Let $x$ be in $K$ and $p \in \partial^- u(x)$, then by semiconvexity

$$u(y) \geq u(x) + p \cdot (y - x) + C|y - x|^2 \quad \forall y \in \Omega.$$

Choosing $y = x + tp/|p|$, we obtain

$$|p| \leq \frac{\mathrm{osc}_\Omega\, u}{t} + C_- t, \qquad 0 < t < \mathrm{dist}(K, \partial\Omega).$$

Minimizing in $t$ we obtain

$$|p| \leq \begin{cases} \dfrac{\mathrm{osc}_\Omega\, u}{\mathrm{dist}(K, \partial\Omega)} + C_-\, \mathrm{dist}(K, \partial\Omega) & \text{if } \sqrt{\frac{\mathrm{osc}_\Omega\, u}{C_-}} \geq \mathrm{dist}(K, \partial\Omega) \\[3mm] \sqrt{C_-\, \mathrm{osc}_\Omega\, u} & \text{if } \sqrt{\frac{\mathrm{osc}_\Omega\, u}{C_-}} \leq \mathrm{dist}(K, \partial\Omega), \end{cases}$$

from which the claim follows.  □

# Appendix B
# A proof of John lemma

In this appendix we give a proof of John lemma, [71]. The proof we give is taken from [70].

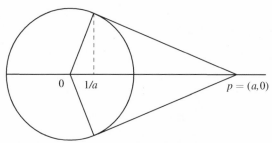

**Figure B.1.** The situation in the proof of John lemma.

**Theorem B.1 (John).** *Let $C \subset \mathbb{R}^n$ be a bounded convex set with non-empty interior. Then there exists a unique ellipsoid $E$ of maximal volume contained in $C$. Moreover this ellipsoid satisfies*

$$E \subset C \subset nE, \tag{B.1}$$

*where the dilation is done with respect to the center of $E$.*

*Proof.* Existence of $E$ is immediate. For the uniqueness just notice that the Minkowski sum of two ellipsoids is still an ellipsoid and use the strict concavity of the map $A \mapsto \det^{1/n}(A)$ on the cone of non-negative symmetric matrices. Let us prove (B.1). Up to an affine transformation we can assume that $E = B$, the unit ball centered at the origin. Let $p$ be the farthest point of $C$, up to a rotation we can assume that $p = (a, 0)$, $a > 0$, see Figure B.1, and our goal is to show that $a \leq n$.

By convexity, the cone $\mathcal{C}$ generated by $B$ and $p$ is contained in $C$. We will show that, if $a > n$ then there exists an ellipsoid with volume strictly larger than $|B|$ contained in $\mathcal{C}$, see Figure B.2.

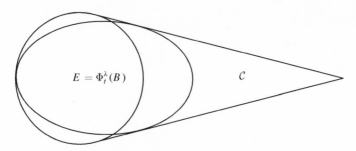

**Figure B.2.** If $a \gg 1$ then there exists an ellipsoid with volume strictly larger than $|B|$ contained in $\mathcal{C}$.

Let us consider, for $\lambda > 0$, the flow of affine maps $\Phi_t^\lambda$ generated by the vectorfield

$$v^\lambda(x_1, x') := \big((x_1 + 1), -\lambda x'\big).$$

An easy computation shows that

$$\Phi_t^\lambda(x_1, x') = \big(e^t(x_1 + t), e^{-\lambda t}x'\big),$$

so that, if $\lambda < 1/(n-1)$,

$$\big|\Phi_t^\lambda(B)\big| = |B|e^{(1-\lambda(n-1))t} > |B| \quad \forall t > 0.$$

We want now to understand when $\Phi_t^\lambda(B) \subset \mathcal{C}$ for small $t$. For this notice that, by similarity, the "straight" part of the cone touches $\partial B$ on the $(n-2)$ sphere

$$S = \big\{x_1 = 1/a, \ |x'|^2 = (a^2 - 1)/a^2\big\},$$

see Figure B.1. To prove that $\Phi_t^\lambda(B)$ is contained in $\mathcal{C}$ for $t$ small, it is enough to show that the vectorfield points inside $\partial B$ in the region where $x_1 \leq 1/a$. This means, recalling that $\nu_{\partial B}(x_1, x') = (x_1, x')$,

$$0 > v^\lambda(x_1, x') \cdot \nu_{\partial B}(x_1, x') = x_1^2 + x_1 - \lambda|x'|^2 \quad (x_1, x') \in \partial B \cap \{x_1 \leq 1/a\}.$$

Since, on $\partial B$, $|x'|^2 = 1 - x_1^2$ the above equation is equivalent to ask that

$$x_1 + x_1^2 < \lambda(1 - x_1^2) \quad \text{for all } -1 \leq x_1 \leq 1/a,$$

which is true if and only if $\lambda > 1/(a-1)$. In conclusion, if $a > n$, we can find a $\lambda$ such that

$$\frac{1}{a-1} < \lambda < \frac{1}{n-1},$$

but then the ellipsoids $\Phi_t^\lambda(B)$ are contained in $\mathcal{C} \subset C$ for small $t$ and have volume strictly greater than the one of $B$, a contradiction. $\qquad \square$

# References

[1] G. ALBERTI and L. AMBROSIO, *A geometric approach to mono-tone functions in* $\mathbb{R}^n$, Math. Z., **230** (1999), 259-316.

[2] L. AMBROSIO, *Transport equation and Cauchy problem for BV vector fields*, Invent. Math., **158** (2004), 227–260.

[3] L. AMBROSIO, *Transport equation and Cauchy problem for non-smooth vector fields*, In: "Calculus of Variations and Non-Linear Partial Differential Equations", Lecture Notes in Mathematics (CIME Series, Cetraro, 2005) **1927**, B. Dacorogna, P. Marcellini (eds.), 2008, 2–41.

[4] L. AMBROSIO, *Transport equation and Cauchy problem for non-smooth vector fields*, In: "Calculus of Variations and Non-linear Partial Differential Equations", Lecture Notes in Math., 1927, Springer, Berlin, 2008, 1–41.

[5] L. AMBROSIO, M. COLOMBO, G. DE PHILIPPIS and A. FIGALLI, *Existence of Eulerian solutions to the semigeostrophic equations in physical space: the 2-dimensional periodic case*, Comm. in Partial Differential Equations, to appear.

[6] L. AMBROSIO, M. COLOMBO, G. DE PHILIPPIS and A. FIGALLI, *A global existence result for the semigeostrophic equations in three dimensional convex domains*, preprint 2012.

[7] L. AMBROSIO, G. CRASTA, V. DE CICCO and G. DE PHILIP-PIS, *A nonautonomous chain rule in* $W^{1,1}$ *and in* $BV$, Manuscripta Math., to appear.

[8] L. AMBROSIO, N. USCO and D. PALLARA, "Functions of Bounded Variation and Free Discontinuity Problems", Oxford Mathematical Monographs, The Clarendon Press, Oxford University Press, 2000.

[9] L. AMBROSIO, G. DE PHILIPPIS and B. KIRCHHEIM, *Regularity of optimal transport maps and partial differential inclusions*, Atti Accad. Naz. Lincei Cl. Sci. Fis. Mat. Natur. Rend. Lincei (9) Mat. Appl. **22**, (2011), 311–336.

[10] L. AMBROSIO, G. DE PHILIPPIS and L. MARTINAZZI, *Gamma-convergence of nonlocal perimeter functionals*, Manuscripta Math. **134**, (2011), 377–403.

[11] L. AMBROSIO and N. GIGLI, *User guide to optimal transport theory*, CIME Lecture Notes in Mathematics, B.Piccoli and F.Poupaud Eds., to appear

[12] L. AMBROSIO, N. GIGLI and G. SAVARÉ, "Gradient Flows in Metric Spaces and in the Space of Probability Measures", Second edition. Lectures in Mathematics ETH Zürich. Birkhäuser Verlag, Basel, 2008.

[13] J.-D. BENAMOU and Y. BRENIER, *Weak solutions for the semigeostrophic equation formulated as a coupled Monge-Ampère/ transport problem*, SIAM J. Appl. Math. **58** (1998), 1450–1461.

[14] M. F. BETTA, F. BROCK, A. MERCALDO and M. R. POSTERARO, *A weighted isoperimetric inequality and applications to symmetrization*, J. of Inequal. & Appl. **4** (1999), 215–240.

[15] L. BRASCO, G. DE PHILIPPIS and B. RUFFINI, *Spectral optimization for the Stekloff–Laplacian: the stability issue* , J. Funct. Anal. **262** (2012), 4675–4710.

[16] Y. BRENIER, *Polar factorization and monotone rearrangement of vector-valued functions*, Comm. Pure Appl. Math. **44** (1991) 375–417.

[17] F. BROCK, *An isoperimetric inequality for eigenvalues of the Stekloff problem*, ZAMM Z. Angew. Math. Mech. **81** (2001), 69–71.

[18] L. CAFFARELLI, *A localization property of viscosity solutions to the Monge-Ampère equation and their strict convexity*, Ann. of Math. (2), **131** (1990), 129–34.

[19] L. CAFFARELLI, *Interior $W^{2,p}$ estimates for solutions of the Monge-Ampère equation*, Ann. of Math. (2), **131** (1990), 135–150.

[20] L. CAFFARELLI, *Some regularity properties of solutions to Monge-Ampère equations*, Comm. Pure Appl. Math., **44** (1991), 965–969.

[21] L. CAFFARELLI, *The regularity of mappings with a convex potential*, J. Amer. Math. Soc. **5** (1992), 99–104.

[22] L. CAFFARELLI, *Boundary regularity of maps with convex potentials*, Comm. Pure Appl. Math. **45** (1992), no. 9, 1141–1151.

[23] L. CAFFARELLI, *A note on the degeneracy of convex solutions to Monge-Ampère equations*, Commun. in Partial Differential Equations **18** (1993), 1213–1217.

[24] L. CAFFARELLI, *Boundary regularity of maps with convex potentials. II*, Ann. of Math. (2), **144** (1996), 453–496.

[25] L. CAFFARELLI and X. CABRÉ, "Fully Nonlinear Elliptic Equations", Amer. Math. Soc., Colloquium publications, Vol. 43, 1995.

[26] L. CAFFARELLI, M. M. GONZÁLES and T. NGUYEN, *A perturbation argument for a Monge-Ampère type equation arising in optimal transportation*, preprint, 2011.

[27] L. CAFFARELLI and C. GUTIERREZ, *Real analysis related to the Monge-Ampère equation*, Trans. Amer. Math. Soc. **348** (1996), 1075–1092.

[28] L. CAFFARELLI and Y. Y. LI, *A Liouville theorem for solutions of the Monge-Ampère equation with periodic data*, Ann. Inst. H. Poincaré Anal. Non Linéaire **21** (2004), 97–120.

[29] L. CAFFARELLI, J.-M. ROQUEJOFFRE and O. SAVIN, *Non-local minimal surfaces*, Comm. Pure Appl. Math. **63** (2010), 1111–1144.

[30] L. CAFFARELLI and E. VALDINOCI, *Regularity properties of non-local minimal surfaces via limiting arguments*, Calc. Var. Partial Differential Equations **41** (2011), 203–240.

[31] F. CHARRO, G. DE PHILIPPIS, A. DI CASTRO and D. MÁXIMO, *On the Aleksandrov-Bakelman-Pucci estimate for the infinity Laplacian*, Calc. Var. Partial Differential Equations, to appear.

[32] D. CORDERO ERAUSQUIN, *Sur le transport de mesures périodiques*, C. R. Acad. Sci. Paris Sér. I Math. **329** (1999), 199–202.

[33] D. CORDERO-ERAUSQUIN, R. J. MCCANN and M. SCHMUCKENSCHLÄGER, *A Riemannian interpolation inequality à la Borell, Brascamp and Lieb*, Invent. Math. **146** (2001), 219–257.

[34] M. CULLEN, "A Mathematical Theory of Large-scale Atmosphere/ocean Flow", Imperial College Press, 2006.

[35] M. CULLEN and M. FELDMAN, *Lagrangian solutions of semi-geostrophic equations in physical space*, SIAM J. Math. Anal. **37** (2006), 1371–1395.

[36] M. CULLEN and W. GANGBO, *A variational approach for the 2-dimensional semi-geostrophic shallow water equations*, Arch. Ration. Mech. Anal. **156** (2001), 241-273.

[37] M. CULLEN and R. J. PURSER, *An extended Lagrangian theory of semi-geostrophic frontogenesis*, J. Atmos. Sci. **41** (1984), 1477–1497.

[38] P. DELANOË and Y. GE, *Regularity of optimal transportation maps on compact, locally nearly spherical, manifolds*, J. Reine Angew. Math. **646** (2010), 65–115.

[39] P. DELANOË and F. ROUVIÈRE, *Positively curved Riemannian locally symmetric spaces are positively squared distance curved*, Canad. J. Math., to appear.

[40] G. DE PHILIPPIS and A. FIGALLI, $W^{2,1}$ *regularity for solutions of the Monge Ampère equation*, Invent. Math., to appear.

[41] G. DE PHILIPPIS and A. FIGALLI, *Second order stability for the Monge-Ampère equation and strong Sobolev convergence of optimal transport maps*, preprint 2012

[42] G. DE PHILIPPIS and A. FIGALLI, *Optimal regularity of the convex envelope*, preprint 2012.

[43] G. DE PHILIPPIS and A. FIGALLI, *Partial regularity of optimal transport maps*, preprint 2012

[44] G. DE PHILIPPIS, A. FIGALLI and O.SAVIN, *A note on interior $W^{2,1+\varepsilon}$ estimates for the Monge Ampère equation*, preprint 2012

[45] G. DE PHILIPPIS and F. MAGGI, *Sharp stability inequalities for the Plateau problem*, preprint 2011.

[46] R. J. DI PERNA and P. L. LIONS, *Ordinary differential equations, transport theory and Sobolev spaces*. Invent. Math. **98** (1989), 511–547.

[47] L. C. EVANS and R. F. GARIEPY, "Measure Theory and Fine Properties of Functions", Studies in Advanced Mathematics. CRC Press, Boca Raton, FL, 1992.

[48] A. FATHI and A. FIGALLI, *Optimal transportation on non-compact manifolds*, Israel J. Math. **175** (2010), 1–59.

[49] H. FEDERER, "Geometric Measure Theory", Springer, 1969.

[50] A. FIGALLI, *Existence, uniqueness, and regularity of optimal transport maps*, SIAM J. Math. Anal. **39** (2007), 126–137.

[51] A. FIGALLI, *Regularity of optimal transport maps [after Ma-Trudinger-Wang and Loeper] (English summary)*, Séminaire Bourbaki. Volume 2008/2009. Exposés 997–1011. *Astérisque No. 332* (2010), Exp. No. 1009, ix, 341–368.

[52] A. FIGALLI, *Regularity properties of optimal maps between nonconvex domains in the plane*, Comm. Partial Differential Equations **35** (2010), 465–479.

[53] A. FIGALLI, *Sobolev regularity for the Monge-Ampère equation, with application to the semigeostrophic equations*, Proceedings of the Conference dedicated to the centenary of L. V. Kantorovich, to appear.

[54] A. FIGALLI and N. GIGLI, *Local semiconvexity of Kantorovich potentials on non-compact manifolds*, ESAIM Control Optim. Calc. Var. **17** (2011), 648–653.

[55] A. FIGALLI and Y. H. KIM, *Partial regularity of Brenier solutions of the Monge-Ampère equation*, Discrete Contin. Dyn. Syst. **28** (2010), 559–565.

[56] A. FIGALLI, Y. H. KIM and R. J. MCCANN, *Hölder continuity and injectivity of optimal maps*, preprint, 2011.

[57] A. FIGALLI, Y. H. KIM and R. J. MCCANN, *Regularity of optimal transport maps on multiple products of spheres*, J. Eur. Math. Soc. (JEMS), to appear.

[58] A. FIGALLI and G. LOEPER, $C^1$ *regularity of solutions of the Monge-Ampère equation for optimal transport in dimension two*, Calc. Var. Partial Differential Equations **35** (2009), 537–550.

[59] A. FIGALLI and L. RIFFORD, *Continuity of optimal transport maps and convexity of injectivity domains on small deformations of* $\mathbb{S}^2$, Comm. Pure Appl. Math. **62** (2009), 1670–1706.

[60] A. FIGALLI, L. RIFFORD and C. VILLANI, *On the Ma-Trudinger-Wang curvature on surfaces*, Calc. Var. Partial Differential Equations **39** (2010), 307–332.

[61] A. FIGALLI, L. RIFFORD and C. VILLANI, *Necessary and sufficient conditions for continuity of optimal transport maps on Riemannian manifolds*, Tohoku Math. J. (2) **63** (2011), 855–876.

[62] A. FIGALLI, L. RIFFORD and C. VILLANI, *Nearly round spheres look convex*, Amer. J. Math. **134** (2012), 109–139.

[63] G. FRANZINA and E. VALDINOCI, "Geometric Analysis of Fractional Phase Transition Interfaces", Springer INdAM Series, to appear.

[64] L. FORZANI and D. MALDONADO, *On geometric characterizations for Monge-Ampère doubling measures*, J. Math. Anal. Appl. **275** (2002), 721–732.

[65] L. FORZANI and D. MALDONADO, *Properties of the solutions to the Monge-Ampère equation*, Nonlinear Anal. **57** (2004), 815–829.

[66] D. GILBARG and N. S. TRUDINGER, "Elliptic Partial Differential Equations of Second Order", reprint of the 1998 edition, Classics in Mathematics. Springer-Verlag, Berlin, 2001.

[67] C. GUTIERREZ, *The Monge-Ampére equation*, Progress in Nonlinear Differential Equations and their Applications, 44. Birkhäuser Boston, Inc., Boston, MA, 2001.

[68] C. GUTIERREZ and Q. HUANG, *Geometric properties of the sections of solutions to the Monge-Ampère equation*, Trans. Amer. Math. Soc., **352** (2000), 4381–4396.

[69] H.-Y. JIAN and X.-J. WANG, *Continuity estimates for the Monge-Ampère equation*, SIAM J. Math. Anal. **39** (2007), 608–626.

[70] R. HOWARD, *A proof of John Ellipsoid Theorem*, Notes, available at http://www.math.sc.edu/ howard/

[71] F. JOHN, *Extremum problems with inequalities as subsidiary conditions*, In Studies and Essays Presented to R. Courant on his 60th

Birthday, January 8, 1948, pages 187-204. Interscience, New York, 1948.

[72] Y. H. KIM, *Counterexamples to continuity of optimal transport maps on positively curved Riemannian manifolds*, Int. Math. Res. Not. IMRN (2008), Art. ID rnn120, 15 pp.

[73] Y. H. KIM and R. J. MCCANN, *Towards the smoothness of optimal maps on Riemannian submersions and Riemannian products (of round spheres in particular)*, J. Reine Angew. Math., to appear.

[74] B. KIRCHHEIM, *Rigidity and Geometry of Microstructures*, Lecture notes, available at
http://www.mis.mpg.de/publications/other-series/ln.html

[75] J. LIU, N.S. TRUDINGER and X.-J. WANG, *Interior $C^{2,\alpha}$ regularity for potential functions in optimal transportation*, Comm. Partial Differential Equations **35** (2010), 165–184.

[76] G. LOEPER, *On the regularity of the polar factorization for time dependent maps*, Calc. Var. Partial Differential Equations, **22** (2005), 343–374.

[77] G. LOEPER, *A fully non-linear version of the incompressible Euler equations: The semi-geostrophic system*, SIAM J. Math. Anal., **38** (2006), 795–823.

[78] G. LOEPER, *On the regularity of solutions of optimal transportation problems*, Acta Math. **202** (2009), 241–283.

[79] G. LOEPER, *Regularity of optimal maps on the sphere: The quadratic cost and the reflector antenna*, Arch. Ration. Mech. Anal. **199** (2011), 269–289.

[80] X. N. MA, N. S. TRUDINGER and X. J. WANG, *Regularity of potential functions of the optimal transportation problem*, Arch. Ration. Mech. Anal. **177** (2005), 151–183.

[81] J. MARSHALL and R. PLUMB, "Atmosphere, Ocean and Climate Dynamics: An Introductory Text", Academic Press, 2008.

[82] R. J. MC CANN, *Existence and uniqueness of monotone measure-preserving maps*, Duke Math. J. **80** (1995), 309–323.

[83] R. J. MCCANN, *Polar factorization of maps on Riemannian manifolds*, Geom. Funct. Anal. **11** (2001), 589–608.

[84] S. MÜLLER, *Variational models for microstructure and phase transitions*, Lecture notes, available at
http://www.mis.mpg.de/publications/other-series/ln.html

[85] R. T. ROCKAFELLAR, *Convex analysis*, Reprint of the 1970 original. Princeton Landmarks in Mathematics. Princeton Paperbacks, Princeton University Press, Princeton, NJ, 1997.

[86] T. SCHMIDT, $W^{2,1+\varepsilon}$ *estimates for the Monge-Ampère equation*, preprint 2012.

[87] G. J. SHUTTS and M. CULLEN, *Parcel stability and its relation to semi-geostrophic theory*, J. Atmos. Sci. **44** (1987) 1318–1330.

[88] E. STEIN, "Harmonic Analysis: Real-variable Methods, Orthogonality, and Oscillatory Integrals", With the assistance of Timothy S. Murphy. Princeton Mathematical Series, 43. Monographs in Harmonic Analysis, III. Princeton University Press, Princeton, NJ, 1993.

[89] J. URBAS, *Regularity of generalized solutions of Monge-Ampère equations*, Math. Z. **197** (1988), 365–393.

[90] J. URBAS, *On the second boundary value problem for equations of Monge-Ampère type*, J. Reine Angew. Math. **487** (1997), 115–124.

[91] N. S. TRUDINGER and X.-J. WANG, On the second boundary value problem for Monge-Ampère type equations and optimal transportation, Ann. Sc. Norm. Super. Pisa Cl. Sci. (5) **8** (2009) 143–174.

[92] N. S. TRUDINGER and X.-J. WANG, *On strict convexity and continuous differentiability of potential functions in optimal transportation*, Arch. Ration. Mech. Anal. **192** (2009), 403–418.

[93] X.-J. WANG, *Some counterexamples to the regularity of Monge-Ampère equations. (English summary)*, Proc. Amer. Math. Soc., **123** (1995), 841–845.

[94] R. WEINSTOCK, *Inequalities for a classical eigenvalue problem*, J. Rational Mech. Anal., **3** (1954), 745–753.

[95] C. VILLANI, "Optimal Transport. Old and New", Grundlehren der Mathematischen Wissenschaften [Fundamental Principles of Mathematical Sciences], 338, Springer-Verlag, Berlin, 2009.

# THESES

This series gathers a selection of outstanding Ph.D. theses defended at the Scuola Normale Superiore since 1992.

## Published volumes

1. F. COSTANTINO, *Shadows and Branched Shadows of 3 and 4-Manifolds*, 2005. ISBN 88-7642-154-8

2. S. FRANCAVIGLIA, *Hyperbolicity Equations for Cusped 3-Manifolds and Volume-Rigidity of Representations*, 2005. ISBN 88-7642-167-x

3. E. SINIBALDI, *Implicit Preconditioned Numerical Schemes for the Simulation of Three-Dimensional Barotropic Flows*, 2007. ISBN 978-88-7642-310-9

4. F. SANTAMBROGIO, *Variational Problems in Transport Theory with Mass Concentration*, 2007. ISBN 978-88-7642-312-3

5. M. R. BAKHTIARI, *Quantum Gases in Quasi-One-Dimensional Arrays*, 2007. ISBN 978-88-7642-319-2

6. T. SERVI, *On the First-Order Theory of Real Exponentiation*, 2008. ISBN 978-88-7642-325-3

7. D. VITTONE, *Submanifolds in Carnot Groups*, 2008. ISBN 978-88-7642-327-7

8. A. FIGALLI, *Optimal Transportation and Action-Minimizing Measures*, 2008. ISBN 978-88-7642-330-7

9. A. SARACCO, *Extension Problems in Complex and CR-Geometry*, 2008. ISBN 978-88-7642-338-3

10. L. MANCA, *Kolmogorov Operators in Spaces of Continuous Functions and Equations for Measures*, 2008. ISBN 978-88-7642-336-9

11. M. LELLI, *Solution Structure and Solution Dynamics in Chiral Ytterbium(III) Complexes*, 2009. ISBN 978-88-7642-349-9

12. G. CRIPPA, *The Flow Associated to Weakly Differentiable Vector Fields*, 2009. ISBN 978-88-7642-340-6

13. F. CALLEGARO, *Cohomology of Finite and Affine Type Artin Groups over Abelian Representations*, 2009. ISBN 978-88-7642-345-1

14. G. DELLA SALA, *Geometric Properties of Non-compact CR Manifolds*, 2009. ISBN 978-88-7642-348-2

15. P. BOITO, *Structured Matrix Based Methods for Approximate Polynomial GCD*, 2011. ISBN: 978-88-7642-380-2; e-ISBN: 978-88-7642-381-9

16. F. POLONI, *Algorithms for Quadratic Matrix and Vector Equations*, 2011. ISBN: 978-88-7642-383-3; e-ISBN: 978-88-7642-384-0

17. G. DE PHILIPPIS, *Regularity of Optimal Transport Maps and Applications*, 2013. ISBN: 978-88-7642-456-4; e-ISBN: 978-88-7642-458-8

## Volumes published earlier

H. Y. FUJITA, *Equations de Navier-Stokes stochastiques non homogènes et applications*, 1992.

G. GAMBERINI, *The minimal supersymmetric standard model and its phenomenological implications*, 1993. ISBN 978-88-7642-274-4

C. DE FABRITIIS, *Actions of Holomorphic Maps on Spaces of Holomorphic Functions*, 1994. ISBN 978-88-7642-275-1

C. PETRONIO, *Standard Spines and 3-Manifolds*, 1995.
ISBN 978-88-7642-256-0

I. DAMIANI, *Untwisted Affine Quantum Algebras: the Highest Coefficient of* det $H_\eta$ *and the Center at Odd Roots of 1*, 1996.
ISBN 978-88-7642-285-0

M. MANETTI, *Degenerations of Algebraic Surfaces and Applications to Moduli Problems*, 1996. ISBN 978-88-7642-277-5

F. CEI, *Search for Neutrinos from Stellar Gravitational Collapse with the MACRO Experiment at Gran Sasso*, 1996. ISBN 978-88-7642-284-3

A. SHLAPUNOV, *Green's Integrals and Their Applications to Elliptic Systems*, 1996. ISBN 978-88-7642-270-6

R. TAURASO, *Periodic Points for Expanding Maps and for Their Extensions*, 1996. ISBN 978-88-7642-271-3

Y. BOZZI, *A study on the activity-dependent expression of neurotrophic factors in the rat visual system*, 1997. ISBN 978-88-7642-272-0

M. L. CHIOFALO, *Screening effects in bipolaron theory and high-temperature superconductivity*, 1997. ISBN 978-88-7642-279-9

D. M. CARLUCCI, *On Spin Glass Theory Beyond Mean Field*, 1998.
ISBN 978-88-7642-276-8

G. LENZI, *The MU-calculus and the Hierarchy Problem*, 1998.
ISBN 978-88-7642-283-6

R. SCOGNAMILLO, *Principal G-bundles and abelian varieties: the Hitchin system*, 1998. ISBN 978-88-7642-281-2

G. ASCOLI, *Biochemical and spectroscopic characterization of CP20, a protein involved in synaptic plasticity mechanism*, 1998. ISBN 978-88-7642-273-7

F. PISTOLESI, *Evolution from BCS Superconductivity to Bose-Einstein Condensation and Infrared Behavior of the Bosonic Limit*, 1998. ISBN 978-88-7642-282-9

L. PILO, *Chern-Simons Field Theory and Invariants of 3-Manifolds*, 1999. ISBN 978-88-7642-278-2

P. ASCHIERI, *On the Geometry of Inhomogeneous Quantum Groups*, 1999. ISBN 978-88-7642-261-4

S. CONTI, *Ground state properties and excitation spectrum of correlated electron systems*, 1999. ISBN 978-88-7642-269-0

G. GAIFFI, *De Concini-Procesi models of arrangements and symmetric group actions*, 1999. ISBN 978-88-7642-289-8

N. DONATO, *Search for neutrino oscillations in a long baseline experiment at the Chooz nuclear reactors*, 1999. ISBN 978-88-7642-288-1

R. CHIRIVÌ, *LS algebras and Schubert varieties*, 2003. ISBN 978-88-7642-287-4

V. MAGNANI, *Elements of Geometric Measure Theory on Sub-Riemannian Groups*, 2003. ISBN 88-7642-152-1

F. M. ROSSI, *A Study on Nerve Growth Factor (NGF) Receptor Expression in the Rat Visual Cortex: Possible Sites and Mechanisms of NGF Action in Cortical Plasticity*, 2004. ISBN 978-88-7642-280-5

G. PINTACUDA, *NMR and NIR-CD of Lanthanide Complexes*, 2004. ISBN 88-7642-143-2

Fotocomposizione "CompoMat" Loc. Braccone, 02040 Configni (RI) Italy
Finito di stampare nel mese di aprile 2013
dalla CSR srl, Via di Pietralata, 157, 00158 Roma